A YOUTHFUL
MAN-O'-WARSMAN

"THREE CHEERS FOR THE STARS AND STRIPES!"
An American 44-gun frigate going into action in 1812.

A YOUTHFUL
MAN-O'-WARSMAN

From the diary of an English lad (a *protégé* of the duchess of the fifth Duke of Marlborough) who served in the British frigate *Macedonian* during her memorable action with the American frigate *United States;* who afterward deserted and entered the American Navy, was recaptured by the British and confined in a South African prison and, on being released, again enlisted in the United States Navy

BY

EDGAR STANTON MACLAY

Author of A History of the United States Navy, A History of American Privateers, Reminiscences of the Old Navy, Life and Adventures of Admiral Philip, Life of Captain Moses Brown, U. S. N.; Editor of the Journal of William Maclay (U. S. Senator from Pennsylvania, 1789-1791), Editor of the Diary of Samuel Maclay (U. S. Senator from Pennsylvania, 1803-1809)

GREENLAWN, N. Y.
NAVY BLUE COMPANY
1910

*he Registered Number of this copy is No.*_____

E 36C
.L4

To the Memory of

Rear Admiral Sir Edward Chichester, R. N.

*As an Expression of American Appreciation for His
Very Handsome Conduct in Manila Bay
May 3 to August 14, 1898*

This Work is Respectfully Dedicated

CONTENTS.

CHAPTER I.

PREFATORY.

CHAPTER II.

WANSTEAD AND BLENHEIM.

CHAPTER III.

ABOARD THE MACEDONIAN.

CHAPTER IV.

LIFE ABOARD A FRIGATE.

CHAPTER V.

LIVELY INCIDENTS ABOARD.

CHAPTER XIII.

SIREN'S LIVELY CRUISE.

CHAPTER XIV.

PRISONERS OF WAR.

CHAPTER XV.

Under the Halter's Shadow.

CHAPTER XVI.

Homeward Bound.

2

CHAPTER XVII.

AGAIN AT BLENHEIM.

LIST OF ILLUSTRATIONS.

CHAPTER I.

PREFATORY.

Too little is known by the American public to-day of the trials, privations and, in many instances, indescribable hardships endured by the enlisted men who so gallantly manned and fought our armed craft in the war of 1812. While too much praise cannot be given to our officers for the splendid work they did while in charge of these ships, the "men behind the guns" who bore the heat of the day and the brunt of the battle seem to have been forgotten.

Not that this neglect of the enlisted man was in any way intentional. It was the logical result of unavoidable conditions. Brave, daring and patriotic as our men-o'-warsmen early in the last century unquestionably were, they were not, as a rule, given to letters. Their education as seamen was superbly practical but seldom academic. They had no official reports to pen and, in view of the cumbersome process of chirography in those days, it is not strange that Jack neglected to record in black and white the noble part he played in our "second war for independence"; and the inevitable result was that the great-hearted American public for nearly a century has remained scarce acquainted with his magnificent achievements.

It is for these reasons that the diary of Samuel Leech, who took an active part as an enlisted man all through this war, becomes especially valuable. In 1843 his diary was privately printed and a few copies were circulated but the volume, long since, has been out of print. In all his extensive researches the writer has discovered only one copy so that the present work may well be considered "new material." To the courtesy of the Hon. William Ward Carruth of Massachusetts the writer is indebted for this copy of Leech's diary.

It is, probably, the only connected narrative in existence of an enlisted man who served in our navy in the war of 1812, and for that reason alone, it should be sacredly preserved. Its value is attested by Rear Admiral Alfred Thayer Mahan, U. S. N., who has inscribed on one of its fly leaves the following: "This book possesses a singular interest from its personal testimony to conditions, once of common knowledge, but of which it is now difficult to obtain specific, authentic description. A. T. Mahan, March 15, 1904."

Samuel Leech was a hereditary servant of the

House of Marlborough, his father having been *valet de chambre* to Lord William Fitzroy, son of the Duke of Grafton; while his mother, for more than thirty years, was one of the trusted and confidential maids of Lady Francis Churchill, duchess of the fifth Duke of Marlborough.

It was through the personal influence of the, then, Lady Francis Spencer that our hero obtained what promised to be an unusually advantageous appointment in the royal navy; and, had it not been for an unfortunate incident (as related in Chapter V) through which Lord Fitzroy— brother of Lady Churchill—was relieved of the command of the *Macedonian*, the subject of this work, undoubtedly, would have remained in His Britannic Majesty's service. As it turned out, Leech was captured in the *Macedonian*, deserted from the royal navy, entered the American service, was recaptured by the English, was confined in a South African prison and, on being released, re-entered the American navy and, eventually, became a respected citizen of the United States. The fact that the present Duke of Marlborough has for his duchess a member of a well-known New York family, gives to Americans a special interest in the truly remarkable career of Samuel Leech.

Another feature of commanding interest in

Leech's life is his graphic and detailed description
(as given in Chapters IX and X) of the naval
battle between the *United States* and *Macedonian*,
fought October 25, 1812. As a specimen of
unaffected, yet vivid, word-picturing it will remain a standard.

This engagement was one of the crucial single-
ship actions of naval history. Taken in connection with the other great frigate actions of this
war, it resulted in the first revolution in the
science of naval warfare in the nineteenth century—just as the duel between the *Monitor* and
Merrimac, half a century later, relegated the
wooden fleets of the world to Rotten Row in
order to give place to ironclads.

When the naval experts of Europe began to
study these frigate actions of 1812 (that between
the *Constitution* and *Guerrière,* the *United States*
and *Macedonian* and the *Constitution* and *Java*)
they, for the first time, appreciated the marked
advance Yankee ingenuity had made in marine
warfare; and remodeled their frigates accordingly—or as the London Times, in its issue of
March 17, 1814, said: "exactly upon the plan
of the large American frigates."

So completely did our 44-gun frigates demon-
strate their superiority over British ships of the
same class that the Admiralty issued a confidential

MODEL OF AN AMERICAN 44-GUN FRIGATE.
Courtesy of the Peabody Academy.

circular, directing commanders of British frigates
to *run away* from the *President, United States*
and *Constitution* as the following extraordinary
command will show: "In the event of one of
his Majesty's frigates under your orders falling
in with one of these ships [the American 44-gun
frigates], his captain should endeavor in the first
instance to secure the retreat of his Majesty's
ship but, if he find that he has an advantage in
sailing, he should endeavor to maneuver and
keep company with her without coming to action
in the hope of falling in with some other of his
Majesty's ships, with whose assistance the enemy
might be attacked with a reasonable hope of
success."

Unfortunately, we have little detail of the
battles between the *Constitution* and *Guerrière*
and the *Constitution* and *Java* but, in that between the *United States* and *Macedonian*, Leech
has given us a battle scene of lasting historical
value—doubly valuable because written from the
enemy's viewpoint—in his vivid narration of the
awfully destructive powers of the American frigate, as compared with those of a similar class in
the English navy.

Aside from his invaluable account of the *United
States-Macedonian* fight, Leech has given us
many side-lights on the career of our navy which

are of popular and historical interest. In short, his narrative is one of unusual lucidity and pertinency on the points touched upon and, on careful examination, the writer is satisfied that Leech's account, in general, may be relied upon. Yet, the writer has taken the liberty of culling such of Leech's statements as might be open to dispute so, it is believed, that the residuum, as it appears in these pages, is founded on the solid rock of well-established fact.

CHAPTER II.

WANSTEAD AND BLENHEIM.

On the nineteenth day of March, in the good year 1798, there was born in Wanstead, England, a boy. Such things had happened in this sedate village (so folk say) before the said "good year 1798" and the parish register bears out the assertion that it has happened frequently since.

His name was Samuel Leech—certainly not a name to arrest attention in these days when blood-letting has long since ceased to be the medical panacea for all ailments. Neither was there anything sufficiently remarkable about the first ten years of the Wanstead boy's life to merit a chronicle. But there was something about his parentage and subsequent career that is of peculiar interest to Americans.

Wanstead, to-day, has been swallowed up in the gigantic metropolitan growth of London but at the time of Samuel's birth it was a lovely suburb, reckoned to be some "seven miles from the city." It was when he was a mere lad that Samuel delighted in wandering through its beautiful park, tossing "crums of comfort" to the timid deer as they grazed under massive oaks that had withstood storms for centuries.

Often did he pass the venerable mansion, seated in the sylvan scene like a queen, to the old parish church with its gorgeous stained-glass windows, to attend Sunday School and, with his fellow scholars primly arranged around the deep-toned organ, wait for the curate to discourse.

But Samuel's greatest delight was the annual Easter hunt in Epping Forest when the youngsters were permitted to chase the deer over hill and dale in hopeless but none the less joyous effort to overtake them. When tired of the "hunt" the boys would wander through the forest, picking flowers, playing games or listening to the sweet-singing birds.

Quite appropriately, Samuel's home was in that part of Wanstead called Nightingale Place, on account of the number of these birds in that vicinity. Those were days to which Samuel, in later years, always looked upon as the one great "sunlight spot" of his life.

Samuel's father was a *valet de chambre* or confidential body servant of Lord William Fitzroy, son of the Duke of Grafton and brother of the Duchess of the fifth Duke of Marlborough. As Mr. Leech died while Samuel was scarcely three

years old, our hero had only a vague recollection
of him.

Indeed, it was while returning from the funeral
that Samuel himself came near losing his life.
Dressed in mourning, in keeping with the occasion, he was coming from the parish church at
Walthamstow, where the services had been held
and, noticing some large boys amusing themselves
by swinging on the rail of a fence, thought he
would try it also. The result was that he lost
his hold and fell into a muddy ditch where he was
almost suffocated before he could be pulled out.

Two years after his father's death, Samuel's
mother became an inmate of the family of Lady
Francis Spencer, daughter of the Duke of Grafton and wife of Lord Spencer who afterward
became Lord Churchill. It was through his
mother's influence that Lord and Lady Churchill
took such an interest in young Leech, seeing to
it that he was placed in a good berth in the royal
navy. The succeeding Duke and his Duchess
arranged for Samuel's return to England in
1841 after his desertion from the English service.

As it was impossible for young Samuel to live
with his mother in her new position he was placed
in the care of an aunt, Mrs. Turner, who was
blessed with a family of twenty-two sons and two
daughters. It was while he lived with his prolific

aunt and numerous cousins that Samuel first got
his idea of going to sea. Most of the Turner boys
were sailors and were constantly going and coming from voyages so that our hero soon had his
mind filled with sea yarns and such stories of
adventure as sailor folk delight in spinning before
credulous landmen.

Seated around the bright fireside of a winter's
evening young Leech drank in the wild stories of
adventure and hair-breadth escapes in unquestion-
ing gulps and he soon came to regard the sailor
as superior to all other beings and the seaman's
life the ideal of human existence. Nor did the
fact that three of the Turner boys died at sea
through hardship and exposure and that two
more went down in the 74-gun ship of the line
Blenheim, when she foundered off the Cape of
Good Hope with seven hundred souls on board,
in the least abate his desire for the sea.

Thus five sturdy young men from one family
perished at sea within the three years of Samuel's
stay at his aunt's home. Had he been superstitious he would have augured evil omen from
the fact that the *Blenheim* was named after the
place where his mother was then living.

When Samuel was about six years old it became inconvenient to have him at the Turner
home and he was placed in charge of a widowed

aunt. This was a sad change for the lad's new
guardian proved very unkind and severe; chas-
tising him for the breakage of a cup or any other
slight accident. Occasionally, Samuel would earn
a few pennies by holding a horse or running errands for neighbors, which his aunt would take
from him as payment for crokery he broke.

One day a smart, jolly tar, fully six feet tall,
knocked at the front door. He said that he had
just returned from that far distant country called
America where he had met a young man named
George Turner, her nephew and Samuel's cousin.
He proceeded to tell many fine stories about
George and finally asked Samuel's aunt if she
would like to see him and if she would know him.

"I don't believe I would know him," said the
widow, "for he has been away so long."

"Well, then," he replied, "I am George
Turner."

The young man had been away eleven years
and after visiting his parents had taken this
method of surprising his aunt. Many were the
yarns he told and many the little gifts and kind
words he bestowed on lonesome, heartsore Samuel. Was it any wonder, then, that the lad came
to regard sailors as the noblest of human kind?

While at this place an uncle from the West
Indies visited the house. He had been one of two

brothers of Samuel's mother who had been educated in Greenwich for the navy. One of them entered the service and, by dint of hard work, finally obtained a commission but soon afterward perished at sea. The other brother had entered the merchant service and, settling in Antigua in the West Indies, accumulated a competence.

One day this uncle took Samuel to London and visited the West India Docks. He was well acquainted with the captains and they paid special attention to our hero—patting him on the head and remarking what a fine sailor lad he would make and asking if he would not like to ship as a cabin boy etc. These flattering attentions served to increase Samuel's desire to go to sea and he returned to Wanstead more dissatisfied than ever with the quiet village life. Soon afterward this uncle went to Trinidad where he died.

It was not long after this that Mrs. Leech, tiring of widowhood, married a Mr. Newman who was a carpenter employed by the Duke of Marlborough and, now having a home of her own, she decided to take Samuel with her. It was a joyful day when this boy mounted the stage coach with his mother and, bidding goodby to his relatives, set out on the long sixty-mile ride to Woodstock.

The tediousness of the journey was relieved

by the antics of a fellow passenger, a sailor, who
cut all sorts of pranks. From spinning yarns he
would execute a hornpipe on the roof of the
coach. Whenever the vehicle arrived at the foot
of a hill he would jump off, run a short distance
and then spring back again with the agility of a
monkey—much to the amusement of the other
passengers and to the no small admiration of our
hero.

At Woodstock our friends left the stage and
covered the remaining distance on foot. Proceeding through the magnificent park of Blenheim they passed under the triumphal arch erected
to the memory of John Churchill, the first Duke
of Marlborough, by his duchess, Sarah Jennings.
Samuel was much impressed by the grandeur of
Blenheim Palace, which was built at public expense and presented to John Churchill for his
victories over the French and Bavarians; but
more especially for his great victory at Blenheim
on the banks of the Danube, August, 1704.

Crossing the park toward Bladen, Samuel was
kindly received by his stepfather. He was a
man in comfortable circumstances, owning the
house in which he lived ; a stone structure, shaded
by a noble apricot tree and surrounded by a
pretty garden.

Samuel found Bladen and the surrounding

3

country quite as beautiful as that at Wanstead.
Well-tended farms, flocks of sheep quietly graz-
ing on the hillsides, expansive fields surrounded
by hawthorn hedges, massive wheatricks and
quaint, old-fashioned farm houses with thatched
roofs met the eye on all sides; while carefully
cultivated gardens and numerous wild flowers
offered gentle, soothing incense.

The people here were very sociable in their
habits and gave the newcomer a hearty greeting.
Once a year they held a great feast called the
Bladen Festival at which they invited all their
friends from surrounding towns. The ceremony
began on a Sunday and for three days eating,
drinking, gossiping and all manner of games were
the order.

Amid such happy scenes time flew rapidly
with our hero. At the age of eleven years he
completed his schooling and was taken into the
Duke of Marlborough's employ as a gardener's
assistant in Blenheim Park. Samuel's early longing for the sea, however, had not left him and
the flame was fanned by tales related by some of
the servants in Blenheim Palace who visited his
home.)

One of the frequent visitors at the Newman
home was a fine, tall fellow, a footman in Blenheim Palace who had been, in his earlier days, a

sailor. He possessed a good voice and whiled
away many an evening with songs, some of
which were in a nautical strain. One of them,
in spite of its somewhat rueful title, " Poor Little
Sailor Boy," especially commended itself to Samuel and he frequently asked for it.

Another visitor was a sergeant in Lord Francis
Spencer's regiment of cavalry and was then attending his Lordship at Blenheim " on duty with
leave of absence." This old soldier also had been
a sailor in his youth and many were the hours
pleasantly beguiled around the Newman fireside
on cold winter evenings listening to his stories
of adventure in foreign parts.

Samuel's mother also showed that she had
inherited some liking for the sea for frequently
she would emphasize these yarns in the lad's
mind by remarking on the noble appearance made
by the ships she had seen when on a visit to
Brighton.

As a result of these converging influences it
soon came to a pass where Samuel could content
himself no longer with the quiet life at Blenheim
and, one day, his mother mentioned the circumstances to Lady Spencer. It just happened at
that time that Lady Spencer's brother, Lord
William Fitzroy, was expecting to command the
frigate *Macedonian* and was at that moment at
Blenheim on a visit, preparing to go to sea.

It will be remembered that Samuel's father, at the time of his death, was Lord Fitzroy's *valet*. Any one understanding the affection (unobtrusive and inconspicuous though it may have been) that exists between hereditary master and servant in old English families, will at once appreciate the strong claim the boy Samuel had on the kind offices of Lord Fitzroy.

So, when Mrs. Newman broached the subject to Lady Spencer, it is not surprising that we find that she took an immediate interest in the boy; not only because of the lad's feudal connection with the family but because she held Mrs. Newman herself in the highest esteem—having had her as a personal attendant many years. At the first opportunity Lady Spencer submitted the case to her brother.

Lord Fitzroy at once sent for Samuel. Trembling in every limb, the boy was ushered into the august presence and to the kindly inquiry if he would like to go to sea, gasped:

" Ye—er—yes, my lord, I would."

After some further questioning the lad was dismissed and it is a tradition in the Newman family (many members of which are living to-day in or near Blenheim) that his Lordship was heard to say:

" I will take that lad under my personal care and see to his future advancement."

Such a high honor from their hereditary lord
and master was sufficient to overcome any scruple
the fond mother had about her boy venturing on
the sea and from that moment it was decided in
the Newman household that Samuel " a sailor
shall be."

CHAPTER III.

ABOARD THE MACEDONIAN.

This great question having at last been definitely settled, namely: that the boy Leech would go to sea, everything was shaped accordingly. Visits of congratulation seemed to be interminable. Many and various were the advices so generously offered. In fact, the whole village seemed to have appointed itself a Committee of One to see that the budding sailor was "professionally" sent off on his new career.

While much of this counsel was honestly intended and, as a rule, painted the future of the boy in glowing colors, there were (it must be confessed) some doubting shrugs of the shoulder and *sotto voce* remarks dropped that: it was not so pleasant aboard a man-of-war after all—even if the chick of a powder-monkey were placed under the protecting wing of the commander. Ominous hints about "flogging" were surreptitiously dropped and other equally harsh punishments were darkly insinuated.

But few of these sobering ideas seemed to have reached the ear of our hero or, if they did, they had lost their effect for so full was he with joy of at last being permitted to get on his favorite element that he was deaf to everything save praise for his new calling.

At last the great day, July 12, 1810, arrived. That it was a "red-letter" day for Bladen it is not necessary to say for was not the youthful seaman the special *protégé* of Lord William Fitzroy, brother of the future Duchess of Marlborough? The whole village turned out in honor of the occasion and amid many honest tears, godspeeds and unnecessary advice, Samuel Leech (attended by his mother, for that good woman was going to see him to the last moment) set out from the Marlborough Arms for London to enter upon his new career.

Instead of going direct to the metropolis, Samuel and his mother paid a short visit to Wanstead where they were very hospitably and affectionately entertained by friends and relatives. Proceeding to London they engaged a boat and were taken down the Thames to Gravesend where they stayed over night—for it was near this place that the *Macedonian* was fitting.

Bright and early on the following morning Samuel, attended by his mother, visited some of the shops in this shipping center and in a short

time, much to the lad's glory, he was rigged in
a " real " sailor suit.

"At last," thought Samuel, " I am a sailor " ;
and he strutted along the streets in a boyish effort
to assume the rolling gait of the true man-of-
warsman, feeling several pegs taller than nature
warranted.

That he should not lack some of the comforts
of life, his good mother purchased for him a chest
filled with wearing apparel and, as her last token
of maternal affection, she gave him a prayer-
book, a Bible and a pack of cards.

Thus equipped, mother and son hired a boat
and, proceeding down the river some two miles,
boarded the frigate *Macedonian* in all the confidence bred of well-assured " influence " with her
commander. Much to their disappointment, Lord
Fitzroy was not aboard so the boy was turned
over to the not too tender mercies of an underling
who took it upon himself to enlist Samuel in
the royal navy " for life."

The lad, however, was inclined to regard this
more as a compliment than a draw-back and,
having his mother with him all day to give advice
about personal deportment and things generally
not in the least pertaining to the sea—as only an
affectionate mother can—he passed the happiest
day of his life. Toward night Mrs. Newman

bade her son an affectionate goodby. The lad leaned far over the rail as he watched the retreating boat, bearing his mother away, and waved adieus to her until the craft disappeared around a bend in the stream; when he lost sight of her—not to see her again until thirty years had rolled past.

The next morning Samuel was put in a "mess," the crew being divided into messes of eight each, who had their meals at one table. This mess proved unfortunate for our hero for it was composed of old tars who did not relish the idea of being so intimately associated with the stripling landlubber.

One of the men, named Hudson, took a special dislike to young Leech and became so persistently abusive in his manner that other members of the mess, out of humanity, advised our hero to change. This is a privilege that was granted in a man-of-war, men being allowed to change around until they find congenial associates. For those unfortunate ones who cannot find desirable messes, a separate table is reserved.

At first, Samuel found it hard to accustom himself to the rough ways of the men about him but, keeping a stout heart, he made honest effort to please and soon had friends. He had the satisfaction of seeing other boys fare worse than himself.

One poor Irish lad, named Billy Garvey, had been seized on shore and was compelled to enter the ship's company. He knew nothing whatever about the sea or sailors and one of his first inquiries on coming aboard was, where he would find his " bed." His messmates told him to inquire of the burly boatswain. That important official looked at the greenhorn a moment and, turning the huge lump of tobacco into another corner of his mouth, asked:

" Have you got a knife? "

" Yes, sir."

" Well, stick it into the softest plank in the ship and take that for a bed."

The poor fellow keenly felt the rudeness for he had been brought up in comfortable circumstances. One day he confided to Samuel:

" When I was at home, I could walk in my father's garden in the morning until the maid would come and say:

" ' William, will you come to your ta, or your coffee-ta or your chocolara-ta? ' But oh! The case is altered now. It's nothing but ' bear a hand, lash and carry.' Oh dear! "

At last, everything being in readiness, the order " Up anchor, ahoy," was given and, for the first

time in his life, Samuel experienced that indescribable thrill of being in a moving ship. Clear-
ing the Thames, the *Macedonian* put into Spithead where she was to take on board about three
hundred soldiers, who were to be transported to
Lisbon to assist the Portuguese in their fight
against Napoleon's army. The soldiers were
stowed away very uncomfortably on the main
deck while their officers messed and bunked with
the officers of the ship.

After a pleasant passage of about a week, the
coast of Portugal was sighted and, taking aboard
a pilot, the *Macedonian* beat about the mouth of
the Tagus (which, at the coast, is nine miles
wide) for nearly a whole day, waiting for a
favorable breeze; for Lisbon was situated some
ten miles up the river and the current was rapid.
At last, favored by a fine breeze, the noble frigate
passed up between the steep, fertile banks of the
Tagus and, passing Half-Moon Battery, shot past
Belem Castle into the port of Lisbon; which, at
that time, was crowded with men-of-war and
transports.

As Samuel stood on the deck of the ship and
looked over the picturesque battlements, cathedral
spires and towers, the city presented a charming
appearance; at least, so he thought. But he
changed his mind when he took a stroll on shore,

through the narrow, ill-kept streets, beset at every hand by beggars ; and he returned to the frigate, satisfied to abide there while she was in port.

It was while at Lisbon that a " great misfortune " befel the *Macedonian*—so her officers declared—for the wardroom steward (Mr. Sanders) deserted. He had long been in the service, was an exceptionally fine provider and knew, to a nicety, how to tickle the palates of his masters. Unlike many men in his profession, Mr. Sanders had carefully saved his prize money, wages, tips etc., and had ˙accumulated a comfortable little fortune.

Not being permitted, by the regulations, to retire from the service he decided to desert and, speaking the Spanish language fluently, he engaged a native boatman to run his boat under the stern of the frigate. Passing through one of the cabin windows, Sanders dropped into the boat and was rowed away—the boatmen concealing him with their flowing garments. It was lucky for Sanders that he was not retaken for, if he had, he would have been subjected to the severest flogging—or hung.

Samuel's exact position on board the frigate was that of servant to the surgeon (and in time of battle he was " powder-monkey," one of the lads who supplied the guns with cartridges) in

which capacity he was compelled to go ashore
many times on errands. On one of these occasions he was shocked by witnessing a brutal
murder, in true Portuguese style. The victim
had aroused the jealousy of a rival, the latter
crept up behind the former and thrust a long
knife, up to its hilt, in his back. It was a
cowardly attack but Samuel soon learned that it
was typical of the natives.

In fact, it soon became well understood that a
calm front was the safest possible protection
against a Portuguese. The Macedonians discovered this on several occasions. At one time
six marines, not understanding the language,
trespassed on the private grounds of the queen.
Some twenty natives rush at them, in a most
ferocious manner, with drawn knives. The Englishmen drew their bayonets and awaited the
onslaught. But, before they came to close
quarters, the natives thought better of their valor
and retreated.

While walking along the streets of Lisbon,
on another occasion, Samuel learned something
about their way of punishing criminals. Noticing
a noisy crowd, he looked up and saw a human
head with a pair of hands nailed to a barber's
pole. On inquiry, he found that they belonged

to a barber who murdered a gentleman he was shaving in order to get a beautiful watch.

It was while the *Macedonian* was at Lisbon that young Leech witnessed a punishment that made a lasting effect on his tender mind. Near where the frigate lay, was anchored an English 74-gun ship of the line. It seems that a sergeant of marines in that ship has especially aroused the anger of those immediately under him by repeated acts of tyranny; and two of the marines determined to take his life.

Waiting for a favorable opportunity, these men one dark night when the deck was comparatively deserted, seized the sergeant and, hurrying him to the side of the ship, threw him into the river. The tide was running swiftly and, as he was securely gagged and bound, he soon perished; no one, save the two marines, being cognizant of the deed. It was not likely that the murder would ever have been found out had it not been for the indiscretion of the marines themselves.

One night, after they had had a day of "liberty" on shore, they came aboard under the influence of liquor and were so loud in discussing the details of the foul act and in congratulating themselves in having rid the ship of the tyrant, that they were overheard by an officer.

A court-martial was convened, the two marines
were tried and found guilty. It was an offense
that admitted of no delay or trifling for it
touched upon the life of discipline in a war ship.
On the following morning, the entire ship's company was assembled on the main deck. The two
criminals, with halters around their necks, were
placed under yard-arms and two guns were fired.
When the smoke had cleared away, two human
bodies were seen dangling at a dizzy height and,
in a few moments, all was again quiet—two lives
had been snuffed out. Only the day before, a
letter had arrived honorably discharging from
the service one of the men executed.

The first Christmas aboard the *Macedonian*
for our hero, tended largely to chill his ardor for
the service. "On this day," said Leech in his
diary, "the men were permitted to have full
swing. Drunkeness ruled the ship. Nearly every
man, with most of the officers, was in a state of
beastly intoxication by night.

"Some were fighting but were so insensibly
drunk that they hardly knew whether they struck
the guns or their opponents. Yonder, a party
was singing bacchanalian songs, such as they
would not have been permitted to do aboard any
ship under normal conditions. All were laughing, cursing, swearing or hallooing. Confusion

reigned in glorious triumph; it was the very chaos of humanity. Had we been at sea, a sudden gale must have proved our destruction. Had we been exposed to sudden attack from an enemy's vessel, we would have fallen an easy prey—just as the poor Hessians at Trenton fell before the well-timed blow struck by Washington, Christmas night, 1776."

CHAPTER IV.

LIFE ABOARD A FRIGATE.

In these days when war ships are built of steel, with every imaginable contrivance of modern invention going into their equipment and armament, much of the old-time romance of sea life is lost. It was a very different condition that Samuel found when he became fairly settled in the frigate *Macedonian*. Here, he discovered an isolated community, cut off for the time being from the rest of the world and governed by a code of regulations peculiar to itself.

He soon learned that every man in that great ship's company had a certain task alotted exclusively to him and, on its proper performance, depended his good standing on the frigate's merit-roll. One set of men (called topmen) was assigned to the duty of handling the sails aloft. They were divided into three divisions namely, fore- main- and mizzen-topmen, according to the number of masts in the ship. It was their duty to ascend their respective masts and to take in, reef or let out sails.

4

Assisting these topmen was a corresponding set of men who handled the sails from the deck. They, also, were divided according to the number of masts and were called forecastle men, waisters (or mainmast men) and the after-guard or those attending to the last (or mizzen) mast in a three-masted ship. They looked after the courses (or lowest sails in a ship), set or took in the jibs, lower studding sails and spanker; and were required to coil up or properly replace all ropes on deck after they had been used.

Another set of men, called scavengers, were required to keep the decks clean; that is, to sweep up and clear away all dirt or refuse from any part of the ship and throw it overboard.

Then came the "boys," among whom Samuel found himself enlisted. They acted mostly as servants for officers and there were from twenty to forty of them in each frigate. The entire ship's company (with the exception of the commander, purser, surgeon and a few other officers and the boys) were divided into two watches, called the larboard (port) and starboard watch, which relieved each other alternately so that when at sea, one watch was constantly on duty.

Every evening the entire ship's company was drilled at the guns. When the drummer beat to quarters, every man and boy hastened to his prescribed station. There were twenty-four guns in each of the broadsides of the ordinary British frigate in those days, eight men and a boy being assigned to each piece: the men to load, fire, sponge and handle the cannon while the boy was to run to and from the gun to the magazine, to secure supplies of ammunition.

Besides the men and boys mentioned, there were from thirty to forty marines (or soldiers who serve aboard ship) in every frigate. They acted as the police, upholding the authority of the officers, standing guard at various points and, in time of battle, some of them were placed in the rigging so as to attack the enemy with their muskets.

The crew slept in hammocks, swung on the berth deck (or that just below the gun or main deck) and, when called to action, they sprang up, dressed, rolled up their hammocks with a rapidity that a landman could hardly believe possible— each hammock being numbered and placed in the bulwark nettings which had a corresponding number; so that, even on the darkest night, the men knew exactly where to place them.

It was some time before Samuel became accustomed to the food aboard ship—it being so different from that he had had on shore. While in port, the men had fresh bread and meat but, at sea, they were confined to salt pork, hard biscuits and pea soup. Once a week they had flour and raisins with which they made a pudding called "plum duff."

While eating, the men were divided into messes of eight, each mess having its cook who drew the allowances, cooked the meals and washed the "kids" or eating utensils. This cook also drew the grog or liquor for the men, which consisted of a gill of rum per man. This was served at noon every day, under the mistaken idea that it made the men stout, hardy and brave. At four o'clock every afternoon every man received half a pint of wine.

In the American service this, worse than useless "grog," was soon done away with and it is related that when the *Constitution* went into battle with the *Cyane* and *Levant*, some officer thought he was doing right by offering our sailors a tub with which to "brace up their courage." The Yankee tars kicked the tub over saying that they needed no "Dutch courage" in entering a fight.

Samuel had not been in the *Macedonian* long before he was called upon to witness that most brutal and degrading of punishments—flogging. (While it was practiced in other European navies of that day with even greater severity than in

the British, it was sufficiently cruel in the latter
to deserve the brand of condemnation. Although
the details of this form of punishment are revolting, it will be necessary to give some description of it, if the reader is to fully appreciate the
dreadful anxieties our hero passed through after
his desertion from the royal navy and subsequent
capture by the English.

One of the *Macedonian's* midshipmen was
named Gale, whom Samuel describes as a "ras-
cally, unprincipled fellow." Finding his handkerchief one day in the possession of a seaman, Gale
accused him of theft—although the tar protested
that he had found it under his hammock; which
was quite possible as the midshipmen often passed
through the berth deck on inspection and other
duties. The case was reported, a court-martial
convened and Captain Carden sentenced the unfortunate seaman to receive three hundred lashes
through the fleet and imprisonment for one year!

To be sure, stealing, in a man-of-war is one
of the gravest petty offenses but, in this case, the
crime was a very long way from fitting the punishment—especially as there was reasonable doubt
of intentional theft.

On the day appointed the prisoner was taken
into the frigate's launch. This boat had been
rigged for the occasion with poles and grating

to which the prisoner, stripped to the waist, was
firmly bound at his wrists and ankles with rope.
The *Macedonian's* surgeon took his place in the
launch, so as to determine when nature had
reached the extreme limit of endurance and a
boat from every ship in the fleet attended and was
connected by a tow-line with the " execution "
launch—so as to give greater humiliation to the
prisoner.

These preliminaries being completed, the crew
of the victim's ship manned the rigging to better
view the proceedings—for the ordeal was designed as a warning for them also. At the word
from the officer in charge, the *Macedonian's*
boatswain, with coat off and sleeves rolled up,
carefully spread out the nine cords of the " cat "
or whip and brought it down with all his strength
on the bare back of the victim. The flesh crept
and reddened. Lash followed lash with nothing
to break the awful silence save the swish of the
nine cords cutting through the air and landing
with a sickening thud on human flesh, or the
agonizing cries of the prisoner.

In order that the blows might be delivered with
undiminished vigor to the last, the boatswain, on
completing one dozen lashes, handed the brutal
instrument to one of his mates; they delivering

BLOOMING

one dozen lashes each. The first sixty lashes were given alongside the *Macedonian,* in conformity with the custom of giving the greatest number of blows alongside the prisoner's ship so that his gory back might strike terror in the crews of the other ships.

By this time the prisoner's back had been lacerated beyond description, the flesh resembling "roasted meat, burned nearly black before a scorching fire," as Samuel described it. His shirt was now thrown over his wounds, the boatswain and mates returned to their ship, all hands were piped down; and as the procession proceeded to the next ship, the drummer beat the Rogue's March.

At the next man-of-war, the crew manned the yards and rigging as before and her boatswain and his mates descended into the launch, cat in hand. Removing his shirt, he revealed the ghastly spectacle to his shipmates aloft. Then they proceeded to deliver one, two or three dozen lashes, according to the number of ships in the fleet.

This horrible drama was to be enacted at the side of every ship, until the three hundred lashes were given. In this case, however, the attending surgeon, at the end of two hundred and twenty blows, pronounced the prisoner unfit to endure any more. Galled, bruised and agonized as he was, he begged them to deliver the remaining, so that he would not again be compelled to pass through the degrading ordeal. His request was denied. Taken aboard the *Macedonian* the surgeon dressed the wounds and, when partially healed (for human flesh could never recover from such mutilation) the other eighty lashes were delivered before the year of imprisonment began.

Thus the mangled wretch was ruined for life, broken in spirit, all sense of self-respect gone—to be for all his remaining days a crawling, servile, cringing slave to the beck and nod of his fellow men; ready, with sullen alacrity, to obey their slightest wish.

This, of course, was a case of extreme severity but the punishment of "flogging through the fleet" was not uncommon in the British or Continental navies of that day. When the United States 32-gun frigate *Essex*, Captain Smith, visited England shortly before the war of 1812, a deserter from an English war ship sought refuge in her. A British officer came on board and made formal demand for him. On being sent below to get his clothes, the deserter approached a carpenter's bench and with one blow,

cut off his left hand with an ax. Picking up the
severed member with his right hand, he returned
to the quarter deck and flung it at the feet of his
captors saying, that before he would again serve
in a British man-of-war he would cut off his left
foot. Horrified at the sight, the lieutenant left
the *Essex* without his prisoner.

It was a peculiar feature of the punishment of
flogging that officers who, at first, sickened and
fainted at the sight, gradually grew indifferent
and, in some instances, acquired a fiendish crav-
ing for it. Not even the tender flesh of the ship's
boys was safe from this brutal ordeal; only, in
their case, *boys* were called upon to handle the
lash instead of men.

Such being the severity of discipline maintained
in the royal navy at this time, it is no wonder
that we find that so many attempted to run
away. With the more desirable men, however,
the British commanders were lenient. While the
Macedonian was under the orders of Lord Fitzroy, a fine sailor named Richard Suttonwood escaped to an English merchant brig. When too
late, Dick found that this brig was laden with
powder for the *Macedonian* and, on the following
day, she ran alongside. Realizing that he was
caught, Dick made the best of a bad case by

boldly going aboard the frigate and surrendering himself—and Lord Fitzroy pardoned him.

"The crew," records Samuel, "were all delighted at his return on account of his lively disposition and ability to sing comic songs. So joyous were we all at his escape from punishment that we insisted on his giving a concert. Seated on a gun, surrounded by scores of sailors, Dick sang a number of favorite songs" to which even some of the officers listened—although, of course, they did not show it.

Lord Fitzroy appreciated the hard life of the men under him and did every thing to make it pleasanter. While at Lisbon a peculiar character was induced to come aboard the *Macedonian* who did much to "'liven up things." "We had just finished breakfast," said Samuel, "when a number our men were seen running, in high glee, toward the main hatchway. Wondering what was going on, I watched their proceedings with curious eye. The cause of their joy soon appeared in the person of a short, round-faced, merry-looking tar who descended the hatchway amid cries of:

"'Hurrah! Here's Happy Jack!'

"As soon as the jovial little man had set his foot on the berth deck he began to sing. It was a song of triumph, of England's naval glories.

Every voice was hushed, all work was brought
to a standstill while the crew gathered round in
groups to listen to his unequalled performance.
Happy Jack succeeded in imparting his joyous
feelings to our people and they parted with him
that night with deep regret."

CHAPTER V.

LIVELY INCIDENTS ABOARD.

Only a few days after the Christmas carousal (as described in Chapter III), word reached the British admiral at Lisbon that nine French frigates were cruising along the western Spanish coast. In a moment all was excitement and confusion as the English war craft, then in that port, hastened to put to sea in search of the enemy. The 74-gun ships-of-the-line *Hannibal*, *Northumberland* and *Caesar*, with the *Macedonian* and a few smaller war ships, at once dropped down the river in pursuit; every man animated with the keenest desire to fight. But after cruising several days in a futile effort to discover the French fleet, the Admiral signaled "Return to port."

On the passage back, the English fell in with a Scotch ship from Greenock which had been reduced, by a succession of gales, to a most perilous condition. Her masts and rudder were gone while numerous leaks were gaining on the pumps. Finding that it was impossible to save the craft, her people were taken off and she was left to sink.

Before gaining port from this short cruise, an incident happened that seriously affected Samuel's future career; indeed, also that of Lord Fitzroy himself. One night while the topsails were being reefed the sailing-master, Mr. Lewis, in a fit of ill-humor, threatened to flog one of the seamen—which, by the regulations, he had no right to do. Lord Fitzroy was a strict disciplinarian—not only with the crew but with his officers—and he would have been the last to have had even the humblest of his ship's company subjected to injustice; not even at the hands of his sailing-master. The latter held a position of responsibility and authority in the ship equal, almost, to that of the commander himself.

Overhearing Mr. Lewis' threat, Lord Fitzroy took him severely to task and the sailing-master so far forgot himself as to enter into a dispute with his superior. The affair reached such a stage that, on their return to Lisbon, the Admiral ordered a court-martial. Unfortunately the sailing-master had considerable influence with naval authorities and, although Lord Fitzroy was perfectly justified in taking the course he did, the court-martial compromised the matter with the result that both officers were cashiered.

This was a bitter disappointment for Lord Fitzroy for he had just entered upon his cherished

profession. He was relieved of his command and
was succeeded, in rapid succession, by Captains
Carson, Waldgrave and John Surman Carden.
The fact that it was so hard to find a commander
of ability equal to Lord Fitzroy, showed, plainly
enough, how well his professional qualities had
commended themselves to the Admiralty.

Among the popular members of the *Macedonian's* ship's company was a negro named
Nugent. He had a fine presence, polite manners
and easy address which had won for him " promotion " to a wardroom servant. As he was an
American, however, and had been unjustly impressed in the British service, he had long kept a
" weather eye " open for an opportunity to escape.

Soon after the return of the squadron to Lisbon, as just narrated, Nugent found his chance
and managed to get aboard an American ship that
was shortly to sail for the other side of the Atlantic. So far, his effort to escape had been
entirely successful for the British officers had not
been able to trace him.

But one unlucky day, while Nugent was standing in full view on the deck of his ship, an officer
on the *Macedonian's* quarter deck, who had the
very professional habit of leveling his spyglass at
any and every thing in sight, happened to bring
Nugent within the field of his vision ; and, recog-

A BOAT PUTTING OFF FROM AN ENGLISH WARSHIP.
From the original painting by R. Westall.

CALIFORNIA

nizing the deserter, sent a boat off which soon
returned with the offender. Punishment for desertion from the British navy, at that time, was
exceedingly brutal. Nugent was placed in irons
until the ship again got to sea but, owing to his
general popularity and favor with the officers, he
got off without the dreaded flogging.

In those days British officers were unscru-
pulous in impressing men into their service, it
frequently happening that citizens of other nations were seized in the streets and hurried off to
British war ships. Of course, these men could
appeal to their consuls and, in some cases, were
released. But, in many instances (such as Nugent's) the pressgang was sent ashore on the eve
of the ship's sailing so that, before word could
reach the consul, the ship was far out to sea—
beyond the hope of recall.

It was while at Lisbon that Samuel himself
came near being an involuntary deserter and,
consequently, incurring the dreadful penalty for
that offense.

One day some of the *Macedonian's* officers
took him ashore to assist them in making purchases for the ship's stores. Proceeding to a
distant part of the city, where the lad never
before had gone, they gave him a small commission to execute. When he endeavored to re-

turn he got lost in the labyrinth of crooked, narrow streets and, not understanding a word of the native language, he failed to get back until the last ship's boat had left.

Here was a predicament, indeed, for our youthful hero! He rightly conjectured that his absence would be construed as wilful desertion and that a police alarm would at once be sent out for his apprehension. If so taken, no excuse of his would be listened to for the local officials would insist on receiving the large reward offered by the Admiralty for the apprehension of deserters from the royal navy; and the *Macedonian's* officers could not pay that reward without recording the punishment in the ship's books.

Almost distracted by the prospect, Samuel set out to find the Fish Market, for he knew that the landing place was near that spot. It was the only bit of Portuguese geography he had the least familiarity with. Addressing, in English, one person and another, he was answered by empty stares and occasionally (from some citizen more intelligent than his fellows) with an " No entender Englis."

Finally, espying a British soldier, he joyfully ran up to him and exclaimed:

" Good luck to you! Do tell me where Fish Market is, for these stupid Portuguese, bad

luck to them, can't understand a word I say; for
it is all ' No entender Englis.' "

The soldier laughed at this exhibition of British temper and very kindly showed the desired
way.

It was now very late at night and few people
were about. Finally, Samuel saw a native boatman but he could not understand what ship the
lad wanted to board. It happened that the *Macedonian,* at this time, had her mainmast out so
that by pointing and holding up two fingers he
managed to make the boatman understand.

On another occasion Samuel was an unwilling
deserter while his ship was in the Bight o' Lisbon. Happening on shore with two other boys,
they overstayed their time and had to remain in
the city all night. Not being provided with
money they were compelled to wander about the
streets until morning.

All that night they were in constant fear of
being apprehended by the local officials as deserters. To guard against this, they practiced a
deception. Samuel was selected to represent a
midshipman for, if there was an officer present,
they could not be arrested for desertion. By
means of a piece of chalk, a stripe was marked
around his collar which, in the uncertain light
of street lamps, made a fairly good imitation of
the silver lace around a midshipman's collar.

5

Getting safely on board the following morning, the boys were separated for examination; to see if they would tell the same story of their absence. That they were in for punishment, they could not doubt for, in Leech's case, at least, this all-night absence had happened twice in a very short period. Fortunately for the boys, one of them happened to be the servant of First Lieutenant David Hope and a favorite. If the lieutenant's boy was flogged, all must be similarly punished so, in order to save the back of his own boy, all were let off with an admonition.

About this time the *Macedonian* was ordered on another cruise and, being short of men, a pressgang (made up of the most loyal men aboard, armed to the teeth) was sent ashore to seize any desirable man they might meet. They returned with the required number. Some of them were Americans but their protection papers were taken from them and destroyed so that proof of the outrage was lost.

Every thing being in readiness, the *Macedonian* put to sea for a cruise in the Bay of Biscay. Scarcely had she been out two days when she encountered a gale in which she nearly foundered. In his diary Samuel says: "We had just finished dinner when a tremendous sea broke over us, pouring down the hatchway, sweeping

the galley of all its half-cooked contents, then
being prepared for the officers' dinner, and covering the berth deck with a perfect flood. It
seemed as if Old Neptune really intended that
wave to sink us in Davy Jones' locker.

"As the water rolled from side to side within
and the rude waves from without beat against
her, our good ship trembled from stem to stern
and seemed like a human being gasping for
breath in a struggle with death. The women
(there were several on board) set up a shriek, a
thing I had never heard them do before. Some
of the men turned pale, others cursed and tried to
say witty things. The officers started, orders ran
along the ship to "Man the chain pumps" and
to cut holes through the berth deck so as to let
the water into the hold. These orders being rapidly obeyed, the ship was freed from her danger.
The confusion of the moment before was followed by laughing and pleasantries. That gale
was long spoken of as one of great danger."

Soon after this narrow escape, the *Macedonian*
gave chase to two West Indiamen. During the
night it blew so hard that it became necessary to
take in the topsails and it was while in the performance of this duty that one of the impressed
sailors (an American named John Thomson) fell
from the yard, struck some part of the ship and
disappeared in the sea—never to rise again.

Leech records that: " He was greatly beloved by his messmates, by whom his death was as severely felt as when a member of a family dies on shore. His loss created a dull and gloomy atmosphere throughout the ship and it was several days before the hands regained their wonted elasticity of mind and appearance."

While on this cruise, one of Samuel's duties was the cleaning of knives, forks, dish-covers etc., for the wardroom. This work devolved on the wardroom boys in succession. One day, having finished his allotted task, the wardroom steward, a quick tempered man from the East Indies, came to Samuel at dinner time to inquire for the knives. Not recollecting, at the moment, where he had put them, the lad made no reply; whereupon the steward pushed Samuel over a sack of bread and, in falling, his head struck the corner of a locker.

As he felt much pain and the blood was flowing freely, Samuel went to Mr. Marsh, the surgeon, who dressed the wound and advised the lad to take care of it.

Without doubt, the cut would have healed speedily had it not been for the freak of a sailor, a few days afterward. It was one of the " sports " of the seamen, while holystoning and washing down the decks of the frigate, to

"souse" the ship's boys. Owing to the injury
on his head and remembering the admonition
given by the surgeon, Samuel had carefully .
avoided this "medicine." This, of course, was
the surest way of courting its application, for
sailor-folk are quick to note any shirking of what
they believe to be each one's due.

Observing that Samuel was trying to avoid his
"dose," an old tar stole up behind him one
morning and poured a bucketful of cold water on
his head. That night Samuel felt violent pains
in his neck, ears and head and, being no better on
the following morning, he was examined by the
surgeon and excused from duty. Continuing to
grow worse, he was sent to the Sick Bay where
he remained several weeks. Although carefully
attended by the surgeon and most kindly treated
by his shipmates (who, rough as they were,
were honestly sorry for him) he continued failing
until his life was despaired of.

With him, in the Sick Bay, was a tough old
negro named Black Tom whose strong frame had
long since been weakened by dissipation. Black
Tom soon died and it was with peculiarly painful
feelings that Samuel watched them sew his body
up in his hammock and heard them read a short
burial service on the deck above. It was not
necessary for any one to tell the lad what "that

splash " in the water was. He heard the ominous sound as Black Tom's body plunged into the deep, and felt sure that the scene would soon be re-enacted over his own remains.

Indeed, he could tell by the lowered voices about him that all hope of his recovery had gone. But, thanks to a strong constitution and discreet living, Samuel gradually recovered and, by the time the frigate returned to Lisbon from this cruise, he was pronounced fit for duty again.

CHAPTER VI.

CAPTAIN CARDEN'S DISCIPLINE.

On return to Lisbon after the cruise in which Samuel Leech so nearly lost his life, as described in the last chapter, our hero found that his master had secured the services of another boy and Leech was temporarily assigned to the duty of messenger for the officers. It was now that Captain John Surman Carden became commander of the frigate. The men fondly believed that he would prove less severe in discipline, for his appearance gave promise of it. In order to propitiate him they called him a "kind, fatherly old gentleman." But their hopes and arts proved of no avail for he soon demonstrated that he was severer than any of his predecessors. The slightest offense was severely punished and the men were soon wishing that they had the really kind-hearted Lord Fitzroy with them again.

Just before sailing from Lisbon on another cruise, the sailing-master's boy ran away. He was caught, flogged and dismissed the service and our hero took his place so that he once more

was housed under the partially protecting wing
of a " master."

Captain Carden, in spite of the severity of his
manner, was a man of refined tastes and one of
his first official acts was to enlist a band of musicians composed of Germans, Italians and Frenchmen. They had been in a French vessel which
had been captured by the Portuguese. As soon
as Carden learned of this he at once engaged
them for the *Macedonian* with the stipulation that
in case of battle they were not to be called on to
fight but were to be stowed away in the cable tier
until " the clouds blew over." Also, they insisted
that they were not to be flogged.

It was fortunate for these men that they made
the stipulations they did for, soon afterward, the
Macedonian was engaged in battle with the
American frigate *United States* and had they
been exposed on deck some of them, undoubtedly,
would have been put, permanently, out of tune.

The career of this band was singular. Captured from the French by the Portuguese, they
enlisted in the British navy. Captured from the
English by the *United States* they entered the
American service and became the second band of
musicians in our navy—the first having been kid-
napped by Captain McNiell at Messina several
years before and carried across the Atlantic, in

spite of their tearful protests that they had not provided for the maintenance of their families during their enforced absence. This first band was being returned to Italy in the *Chesapeake* when that ship was captured by the British frigate *Leopard* in 1807, and again they were drawn into sharps and flats—from all of which we can readily believe that the lot of the early sea-musician was not a happy one.

During the cruise on which the *Macedonian* had now entered, these musicians played whenever Captain Carden dined and when the wardroom officers messed they played before the door of that sanctum; except on Sundays when Carden was in the habit of honoring the wardroom with his presence, the band accompanying him. On the whole, the crew was much pleased to have the "artists" with them for they enlivened the monotony of the cruise and formed a shining target for the amateur musicians among the sailors to hurl their scornful criticisms at.

A few days after leaving port an incident occurred which will be described in Samuel's own words. "The thrilling cry 'A man overboard!' ran through the ship with electrical effect. It was followed by:

"'Heave out a rope!' then by still another:

"'Cut away the life buoy!' Then came the order:

" 'Lower a boat!'

" Notwithstanding the rapidity of these commands and the confusion occasioned by the anticipated loss of a man, they were rapidly obeyed. The ship was hove-to. By that time, however, the cause of all this excitement was at a considerable distance from the ship. He was a poor Swede named Logholm who, while engaged in lashing the larboard [port] anchor stock, lost his hold and fell into the sea.

" He could not swim but somehow managed to keep afloat until the boat reached him, when he began to sink. The man at the bow ran his boat-hook down and caught the drowning man by his clothes. The cloth giving way, the man lost his hold and the Swede once more sank. Again the active bowman ran the hook down, leaning far over the side. Fortunately he got hold of the dying man's shirt collar. Dripping and apparently lifeless, he was drawn into the boat. He was soon under the surgeon's care, whose skill restored him to life." Poor Logholm! He had escaped death from drowning to await a more dreadful fate in battle, as we shall soon see.

Awaking one morning, soon after this, Samuel discovered that he had had the narrowest escape of his life from getting the lash. He learned that, during the night, a strange vessel had approached and believing her to be an armed enemy, the entire ship's company had been called to quarters. The drums beat, the bugle sounded to arms, the great guns were got in readiness for action, the battle lanterns lighted while the officers and men mustered at each division. When all was in readiness, it was discovered that the stranger was a harmless merchantman; upon which the off-watch returned to their hammocks. Samuel had slept soundly all through the bustle and confusion and, luckily for him, his absence from the assembly for action was not noticed.

Arriving at Madeira, the Portuguese boy who had taken Samuel's place as servant to the surgeon, was dismissed as being unfit for Anglo-Saxon company and it is a high compliment to Samuel's character that the surgeon made strong (though futile) efforts to have our hero again for his needs.

From Madeira the *Macedonian* proceeded to St. Michael's when one of the women on board (wife of one of the crew) gave birth to a fine, plump boy. This happy incident was quickly followed by another birth which tended greatly to relieve the monotony of the cruise. Apparently, Captain Carden did not relish the idea of having his gallant frigate turned into a nursery and, on the return of that ship to Lisbon in a few days, he ordered all the women aboard to enter a vessel about to sail for England.

It was while the *Macedonian* was in Lisbon on this trip that our old friend, Bob Hammond, got into trouble again. While below deck, one day, he was vexed by the taunts of a shipmate and, raising his huge fist, aimed a blow which, instead of striking the offender landed on another seaman. Bob was too angry to apologize and only remarked:

" I have killed two birds with one stone."

Fighting among the crew is a serious offense aboard a war ship and the next morning Bob was ordered before Captain Carden and asked if he had struck the man. Unhesitatingly Bob replied that he had and was glad of it. Two dozen lashes were immediately laid on him and, being taken down from the grating the Captain questioned him again. But Bob only replied that:

" The man who reported me is a scoundrel "— Bob used a stronger word.

For this, the Irishman got another dozen lashes. All the strokes were received without eliciting the slightest groan or twitch from the victim and, thoroughly discouraged, Captain Carden sent him below.

One of the finest sailors in the *Macedonian*— a man popular not only with the crew but with the

officers—named Jack Sadler, growing weary of
the service, determined to desert. One night he
lowered himself over the side into the river and
began swimming toward the shore. As it was
not very dark, he was discovered and the sentry
fired at him but without effect. Then a boat was
lowered and went in chase. Jack was soon overtaken, when the officer in charge said:

"Well, Mr. Sadler, you thought you had got
away, did you?" to which Jack replied:

"You are not so sure that you have me now."

And with that he sprang into the river and
would have escaped had not a boat from another
war ship headed him off; for he was an expert
swimmer and could remain under water a long
time. On account of his popularity he was let
off with three dozen lashes—a remarkably light
punishment for desertion.

One of Sadler's "besetting sins" was that of
drunkenness and a few days after this attempted
escape he applied himself lustily to the bottle—
with the usual result—and was placed in irons.
Jack was Bob Hammond's messmate and the latter, observing his "chum's" condition, instantly
became so sympathetic that he got drunk too—
so that he could "be with him."

While in this condition, Bob purposely placed
himself in the way of the officers and in a short

time had the satisfaction of sharing Jack's irons where the sympathetic souls were soon pouring their commiserations into each others ears. The united effect of their distressed feelings was remarkable for in a short time they began singing and throughout that live-long night they kept up such a yelling and hallooing that not one of the great ship's company could sleep—especially the officers whose rooms were near the " brig "; and they were too dignified to order a change.

When the culprits were hauled up for punishment the following morning, Captain Carden turned to Jack and said:

" Well, Mr. Sadler, you were drunk, were you last night? "

" I was, sir," replied the offender. Jack's offense merited severe handling but he was a favorite of the commander and Captain Carden wished to be lenient. So he said:

" Do you feel sorry for it, sir? "

" I do, sir."

" Will you try to keep sober if I forgive you? " continued Captain Carden.

" I will try, sir," was the reply.

" Then, sir, I forgive you."

Then turning with great severity to Bob Hammond, the British commander said:

" Well, Mr. Hammond, you got drunk, did you, sir? "

Bob shrugged his shoulders and shifting the enormous quid of tobacco to a convenient position in his mouth, replied:

"I can't say but what I had a horn of malt, sir."

In a voice of thunder Carden asked: "A horn of malt? You rascal! What do you call a horn of malt?"

Bob shifted the weight of his body to the other leg and, giving a nautical hitch to his trousers, drolly said:

"When I was in Bengal, Madras and Batavia I used to get some stuff called arrack—we used to call it a horn of malt; but what I had yesterday was some good rum."

While delivering this explanation Bob's manner was so exquisitely ludicrous that both the officers and men burst out laughing. Captain Carden was confused but recovering himself said to First Lieutenant Hope:

"Put that rascal in irons! It's of no use to flog him."

One of Captain Carden's hobbies was to pick out only the finest seamen for his ship and to make room for them he managed to weed out the drones by giving them every opportunity to desert. Once in a while he would call on the men he wished to lose, to go ashore and "cut stuff to make brooms of," as he significantly expressed it. These men soon came to be known as the "Broomers" and it was generally understood that, if they did not return to the ship, no special effort would be made to capture them.

Now, while Bob was an excellent seaman when sober, Captain Carden was fearful of the effect his influence would have on other members of the ship's company and one day, when the "Broomers" were called, Carden said to Hammond in a very knowing manner:

"Mr. Hammond, you may go with these fellows to cut broom stuff." Bob took the hint, replying:

"Ay, ay, sir, and I will cut a long handle to it."

True to his word, he "cut a long handle to it" for he never appeared on the *Macedonian's* deck again.

CHAPTER VII.

FORESHADOWING A GREAT BATTLE.

About this time, 1811, the prevailing topic of conversation among British sailors was the probability of war with the United States and it will prove interesting to observe the confidence generally expressed of the easy victories England would have, at least, on the sea.

Down to this time Samuel had been a loyal Briton and he well expressed the sentiment of his shipmates when he said: "The prevailing feeling through the fleet was that of confidence in our own success and of contempt for the inferior naval force of our anticipated enemy. Every man, and especially the officers, predicted, as his eye glanced proudly on the fine fleet which was anchored off Lisbon, a speedy and successful issue of the approaching conflict."

Shortly after this the *Macedonian* received orders to carry dispatches to Norfolk, Virginia, and she put to sea accordingly. It was then in the "dead of winter" and young Leech, as the frigate neared her haven, began to feel the sharp cold of the American climate. It was the first

6

time he had experienced really cold weather and
he listened with unusual interest to the "cold
weather" yarns the sailors began to spin.) One
story particularly impressed itself on his mind.

It was that about a tyrannical lieutenant who
delighted in imposing extra tasks on the crew.
Although it was never his watch on deck at the
hour holystoning was done, he managed to appear before the task was completed and made the
men do it over again. This, in a severe climate,
was a great hardship so that many a curse or
prayer was offered that he might be taken else-
where.

One morning, the weather being unusually
cold, the men sprang to their work hoping to
finish holystoning the decks before their tormenter came up. They had just completed the
task when the lieutenant appeared and angrily
ordered them to do it all over again. With muttered imprecations they got down to the wet
decks on their knees, earnestly hoping that
he might never again appear on deck alive.
Strangely enough their wish was granted for the
officer was taken sick and died the next day.

Dropping anchor in Hampton Roads, the frigate was made snug for a short stay in American
waters. Samuel Leech records: "The sound of
our own familiar tongue from strangers was very

agreeable after being so long accustomed to Portuguese lingo and a thrill of home remembrances shot through our hearts as the American pilot, on stepping on board said : ' It's very cold.' "

"While in Hampton Roads we fared well. Boats were alongside every day with plenty of beef and pork, which was declared by universal consent to be infinitely superior to that we had obtained in Portugal. Our men said that Yankee pork would swell in the pot, which they very sagely accounted for on the supposition that the pigs were killed in the full of the moon."

The principal drawback to the enjoyment of the stay at Norfolk was the denial of liberty to go ashore. The strictest care was taken to prevent communication with the land, either personally or by letter lest the men would desert or might learn of the impending war. Many of the crew were Americans wrongfully impressed.

Speaking of the interchange of visits between the officers and Americans, Samuel touches on one of the historical episodes in the American navy. He said: "Our officers never enjoyed themselves better than during our stay at this port. Besides feasting among themselves on the fine, fat beef, geese and turkeys which came alongside in abundance, they exchanged visits with Commodore Decatur and his officers of the frigate

United States, then lying at Norfolk. I remember hearing Commodore Decatur and the captain of the *Macedonian* joking about taking each other's ships in case of war."

From Captain Mackenzie's Life of Decatur we get the details of this conversation which, as will soon be seen, was prophetic—though, according to another account it took place in Lisbon instead of at Norfolk. "Carden particularly pointed out the inefficiency of the 24-pounders on the main deck of the *United States* and said that they could not be handled with ease and rapidity in action and that long 18-pounders would do as much execution and were as heavy as experience had proved that a frigate ought to carry. 'Besides, Decatur,' said Carden, 'though your ships may be good enough, and you are a clever set of fellows, what practice have you had in war? There's the rub.'"

After a quick run across the Atlantic, the *Macedonian* arrived at her old quarters in Lisbon where, much to the joy of all on board, orders were found awaiting her to convoy a fleet of merchantmen home. It was now over two years since Samuel had seen Merry England and he describes the departure from Lisbon in the following graphic style: "One morning a gun was fired to give notice to our convoy to get under

ENGLISH NAVAL OFFICERS DINING.
From the original painting by T. Davidson.

weigh.   Immediately the harbor was alive with
noise and activity.   The song of the sailors weighing anchor, the creaking of pullies, the flapping
of sails, the loud, gruff voices of the officers and
the splashing of the waters created what to us,
now that we were homeward bound, was a sweet
harmony of sounds.

"Amid all this animation, our own stately
frigate spread her bellying sails to a light but
favoring breeze.   With colors flying, our band
playing lively airs and the captain with his speaking trumpet urging the lagging merchant ships to
more activity, we passed gaily through the fleet
consigned to our care.   In this gallant style we
scudded past the straggling ruins of old Lisbon,
which still bore marks of the earthquake that
destroyed it.   Very soon the merry fishermen,
who abound in the Tagus, were far at our stern.
Next we glided past the tall granite pinnacles of
towering Mount Cintra.   The highlands passed
from our vision like scenes in a panorama and in
a few hours, instead of the companionship of the
large flock of seagulls, which hover over this
river, we were attended, here and there, by only
one of these restless wanderers of the deep.
Now we were fairly at sea and were enjoying the
rare luxury of fond anticipations.   Visions of
many an old fireside, of many an humble hearth-

stone, poor, but precious, flitted through the
minds of many of our crew that night. Hardships and severe discipline were for the time forgotten in dreams of hope."

After a pleasant passage of a few days the
beautiful shores of Old England greeted the
longing eyes of the *Macedonian's* crew and in a
short time the noble frigate was snugly anchored
in the harbor of Plymouth. But now Samuel experienced another (and perhaps the most trying
hardship) of the sailor's life. Lest the men might
desert, they were not permitted to go ashore although the ship was under orders to undergo
thorough repairs which necessitated a long stay
in that port.

To our hero this was a bitter disappointment,
for he had fondly counted on paying a short visit
to Bladen and again seeing his mother. As a
special favor he was permitted to go ashore with
another boy and this delicious taste of *terra firma*
is described by the lad in the following words:
" One fine Sabbath morning I went ashore with a
messmate who lived in Plymouth and in company
with some children we wandered into the fields
where the merry notes of the numerous birds, the
rich perfume of the blooming trees, the tall green
hedges and the modest primroses, cowslips and
violets, which adorned the banks by the roadside,
filled me with inexpressible delight."

True to the pious teachings of his mother,
Samuel did not abuse the privilege of going
ashore, as some of the other boys in the *Macedonian* did. He kept away from the alehouses
and at sunset returned aboard and reported himself to the officer of the deck. The latter seemed
surprised to find the youngster sober and dismissed him with a kind word.

About the same time some of the other lads
returned from liberty in a condition that showed,
too plainly, that they had been over indulging in
liquor. They loudly berated Samuel for being
so " white-livered and unseamanlike " as not to
get drunk, declaring that he was fit only for the
company of babies and womenfolk. On the following morning, however, these same boys were
summoned to the grating and were soundly
flogged for their conduct; so that, after all, Samuel had the laugh on them.

While at Plymouth the *Macedonian* was taken
into one of the magnificent drydocks of that port
and was thoroughly overhauled. New rigging
was rove, alterations made and, being repainted
inside and out, she looked like a new ship.

Finally, everything being in readiness, the frigate made a quick run over to the French port
of Rochelle from which place she proceeded to
Brest where a formidable English fleet was found

blockading the port. A superior French fleet was in this harbor and it was the plan of the British commander-in-chief to lure it out to a general engagement.

Our hero records: "With all our maneuvering, we could not entice them from their snug berth in the harbor, where they were safely moored, defended by a heavy fort and by a chain crossing the harbor. . . . . Sometimes we sent a frigate or two as near their fort as we dared. At other times the whole fleet would get under weigh and stand out to sea—but without success. Once in a while their frigates would creep outside the forts, when we gave them chase but seldom went beyond the exchange of a few harmless shots."

Returning to Plymouth and then again making for the coast of France, the *Macedonian* had nearly accomplished the distance of the return passage when, one day, the man at the masthead cried out:

"Sail, ho!"

"Where away?" inquired the officer of the deck. Being informed, the officer asked:

"What does she look like?"

"She looks small; I cannot tell, sir."

Waiting a few minutes, until the stranger was nearer, the officer again hailed:

"Masthead, there! What does she look like?"

"She looks like a small sail boat, sir."

So small, indeed, was this sail boat that it was not until the frigate had almost run her down that she could be seen from the deck, when it was found that it contained a man and a boy. That such a frail craft should be alone in these dangerous waters was, in itself, enough to arouse suspicion and running alongside the English took them aboard. They proved to be two French prisoners who had escaped from a British prison and, having stolen a skiff, had ventured on this voyage for liberty.

"Poor fellows," said Samuel. "They looked sadly disappointed in finding themselves again in British hands when in sight of their own sunny France. . . . . I am sure we all would have been glad to have missed them."

Returning to the blockading fleet off Brest, the *Macedonian* entered upon a service that was more "active" than that she had previously engaged in. Finding that the French would not come out, boat expeditions were organized and went out at night to pick up whatever might come in the way.

One night a more formidable expedition than usual was under way and Samuel records that "the oars were muffled, the boats' crews increased and every man was armed to the teeth. The cots

were got in readiness in case any of the party
came back wounded. . . . . Notwithstanding these
omnious preparations, the brave fellows went off
in fine spirits as if they had been going ashore
on liberty. We had no tidings of this adventure
until morning when I was startled by hearing
three cheers from the watch on deck which were
answered by a party that seemed to be approaching us.

"I ran on deck just as our men came alongside with their bloodless prize, a lugger, laden
with French brandy, wine and castile soap. As
this was our first prize we christened her *Young
Macedonian.* Before sending her to England for
condemnation, some of our people replenished
their empty bottles with the brandy.

A MIDNIGHT BOAT ATTACK.
From the original painting by R. Jack.

CHAPTER VIII.

A MOMENTOUS NAVAL BATTLE.

We now come to that part of Samuel Leech's diary which forms one of the most important documents in our naval history and, in consideration of its great value, it will be given in the hero's own words.

The expression "most important" has been used advisedly. There have been more spectacular and more dramatic actions between single warships in the nineteenth century but none (with one exception—that between the *Monitor* and *Merrimac*) has exerted such influence in revolutionizing the science of sea fighting and naval architecture as that group of single-frigate battles in 1812, namely: that between the *Constitution* and *Guerrière*, the *United States* and *Macedonian* and the *Constitution* and *Java*.

Unfortunately we have but scanty details of the first and last mentioned engagements but in the second we have the fullest account in Leech's diary; and it may, unhesitatingly, be accepted as being typical of the three. Leech's account is doubly valuable inasmuch as it is given from

the enemy's view-point; therefore, unbiased in its
acknowledgment of the marvelous accuracy of
American gunnery and the general efficiency
of the early American-built frigate.

It is undeniable that the direct result of this
group of single-frigate actions was the first
revolutionizing of the fleets of the world in the
nineteenth century and the change from European
traditions and methods in naval architecture to
American ideals—just as the battle between the
Monitor and *Merrimac* resulted in the relegation
of wooden war craft to the Rotten Row, to make
room for ironclads.

In order that we may more intelligently follow
the momentous sea fight in which the *Macedonian*
was soon to engage, a few words descriptive of
the opposing ships will be necessary.

The *United States* belonged to that famous
group of frigates—having for her sisters the
Constitution and *President*—built at the close of
the eighteenth century. These ships embodied
many new ideas in naval construction which were
unsparingly criticised by European experts as
" rash innovations," " unprofessional vagaries "
and " visionary schemes." So deeply rooted was
English conservatism in this matter that even
Captain Carden himself, after visiting the frigate
United States (the ship he was fated to fight)

and having thoroughly inspected her, frankly declared his belief that the *Macedonian* was more than her match.

Aside from many details of minor importance, the main difference between the class of large American and British frigates at that day was that the former carried 24-pounders to the latter's 18-pounders on the main deck, and 42-pounders to their 32-pounders on the quarterdeck and forecastle.

But it was just this superiority of weight that Englishmen declared would work detrimentally. They insisted that 24- and 42-pounders could not be handled as efficiently in the heat of action as 18- and 32-pounders.

As a matter of fact, the American 44-gun frigate *was* overweighted (carrying fifty-five guns to the English forty-nine) and the experience of the first three frigate actions proved it; with the result that before hostilities ceased our number of guns had been reduced to fifty-one. It was with this reduced armament that the *Constitution* captured with marvelous immunity to herself the *Cyane* and *Levant* whose combined armaments made fifty-five guns with a total shot weight of 754 pounds to the broadside as opposed to the 644 pounds in the Yankee. To a limited extent, then, we find Captain Carden's remark about the overweight of our frigates well founded and that the *Constitution, United States* and *President* were more formidable with the reduction.

It was fashionable in European court circles, in those days, to refer to anything American in terms of derision and contempt and our gallant little navy bore the brunt of these unkindnesses. At the outbreak of the war of 1812, English newspapers spoke of our frigates as "bundles of fir planks, flying a gridiron flag" while the London Statesman, in its issue early in June, 1812, solemnly declared: "America certainly cannot pretend to wage war with us; she has not the navy to do it with."

Before the war was over, however, England, in self defense, was compelled to follow our models: as will be seen in the following extract from the London Times of March 17, 1814: "Sir G. Collier was to sail yesterday from Portsmouth for the American station in the *Leander*, 54. This ship has been built and fitted out exactly upon the plan of the large American frigates."

Like all of its European contemporaries, The Thunderer expressed unlimited confidence in the ability of the British navy to "sweep the contemptible flag of the United States from the seas" and to "reduce our marine arsenals to a heap of ruins within six months" but, when it heard of

the loss of the *Macedonian*, so quickly following
that of the *Guerrière*, it exclaimed:

" In the name of God, what was done with this
immense superiority of force? "

Summing up the results of the naval war of
1812 the London Times, in its issue of December
30, 1814, said: " We have retired from the
combat with the stripes yet bleeding on our
backs. Even yet, however, if we could but close
the war with some great naval triumph the repu-
tation of our maritime greatness might be partially restored. But to say that it has not hitherto
suffered in the estimation of all Europe and, what
is worse, of America herself, is to belie commonsense and universal experience.

" ' Two or three of our ships have struck to a
force vastly inferior.' No! Not two or three
but many on the ocean and whole squadrons on
the lakes; and the numbers are to be viewed with
relation to the comparative magnitude of the
two navies. Scarcely is there an American ship
of war which has not to boast a victory over the
British flag; scarcely one British ship in thirty or
forty that has beaten an American. With the
bravest seamen and the most powerful navy in the
world, we retire from the contest when the bal-
ance of defeat is so heavy against us "—and these
extraordinary admissions were penned before the

editor had learned of the capture of the *Cyane*
and *Levant* by the *Constitution*, the disabling of
the *Endymion* by the *President*, the capture of
the *Nautilus* by the *Peacock*, the disastrous action
on Lake Borgne, the repulse of the British boat
attack on the *General Armstrong* in Fayal, or the
brilliant victory of the *Hornet* over the *Penguin!*

Such being the peculiarly discouraging conditions under which our little navy entered into the
naval struggle of 1812-'15, it is with special interest that we follow the graphic details Samuel
Leech has given us relative to the second great
frigate action of that war.

That the officers and crew of the *Macedonian*
were thoroughly imbued with that overweening
pride in England's naval prowess, so prevalent at
that date, is shown in a quotation from Leech's
diary which bore on a rumor that war was immi-
nent between Great Britain and the United States
in 1811. He said: "Every man, and especially
the officers [in the *Macedonian*], predicted, as
his eye glanced proudly on the fine fleet which
was anchored off Lisbon, a speedy and successful
issue to the approaching conflict. The prevailing
feeling through the whole fleet was that of confidence in our own success and contempt for the
inferior naval force of our anticipated enemies."

And why should not "every man, and especially the officers " in the *Macedonian* so " predict " when their ship was one of the newest and finest products of British naval construction, whereas, every one of her possible Yankee rivals of the same class were " old tubs " built in the preceding century?

CHAPTER IX.

IN BATTLE'S AWFUL DIN.

(*Leech's own narrative.*)

At Plymouth we heard some vague rumors of
a declaration of war against the United States.
More than this, we could not learn since the
utmost care was taken to prevent our being fully
informed. The reason of this secrecy was, probably, because we had several Americans in our
crew, most of whom were impressed men. These
men, had they been certain that war had broken
out, would have given themselves up as prisoners
of war and claimed exemption from that unjust
service which compelled them to fight against
their country.

This was a privilege which the magnanimity
of the officers ought to have offered them. They
had already perpetrated a grievous wrong upon
them in impressing them. It was adding cruelty
to injustice to compel their service against their
own nation.

Leaving Plymouth we next anchored, for a
brief space, at Torbay, a small port in the British

Channel. We were ordered thence to convoy a
huge East India merchant vessel, much larger
than our frigate and having five hundred troops
on board, bound to the East Indies with money
to pay the troops there.

We set sail in a tremendous gale of wind.
Both ships stopped two days at Madeira to take
in wine and a few other articles. After leaving
this island, we kept her company two days more
and then, according to orders, having wished her
success, we left her to pursue her voyage while
we returned to finish our cruise.

Though without any positive information, we
now felt pretty certain that our Government was
at war with America. Among other things our
captain appeared more anxious than usual. He
was on deck almost all the time. The lookout
aloft was more rigidly observed and every little
while the cry:

"Masthead, there!" arrested our attention.

It is customary in men-of-war to keep men at
the fore and mainmast heads, whose duty it is
to give notice of every new object that may appear. They are stationed at the royal yards, if
those are up but, if not, on the topgallant yards.
At night a lookout is kept on the foreyard only.

Thus we passed several days, the captain running up and down, constantly hailing the man

at the masthead. Early in the morning he began his charge:

"Keep a good lookout" and continued to repeat it until night. Indeed he seemed almost crazy with some pressing anxiety.

The men felt that there was something anticipated of which they were ignorant and had the captain heard all their remarks upon his conduct he would not have felt highly flattered. Still, everything went on as usual. The day was spent in the ordinary duties of man-of-war life at sea and in the evening in telling stories of things most rare and wonderful—for your genuine old tar is an adept in spinning yarns.

To this yarn-spinning was added the most humorous singing, sometimes dashed with a streak of the pathetic which, I assure you, was most touching; especially in one very plaintive melody with a chorus beginning with:

Now if our ship should be cast away,
It would be our lot to see Old England no more.

This song made rather a melancholy impression on my mind and gave rise to a sort of presentiment that the *Macedonian* would never return home again. The presence of a shark following our frigate for several days, with its attendant pilot fish, tended to strengthen this prevalent

idea. [That shark was destined to be kept exceedingly busy for the next few days.—E. S. M.]

The Sabbath [October 25, 1812] came and
brought with it a stiff breeze. We usually made
a sort of a holiday of this sacred day. After
breakfast it was common to muster the entire
crew on the spar deck, dressed as the fancy of the
captain might dictate: sometimes in blue jackets
and white trousers or blue jackets and blue
trousers; at other times in blue jackets, scarlet
vests and blue or white trousers with our bright
anchor buttons glancing in the sun and our black
glossy hats ornamented with black ribbons and
with the name of our ship painted on them.

After muster we frequently had church service,
read by the captain. The rest of the day was
devoted to idleness. But we were destined to
spend this Sabbath in a very different manner.

We had scarcely finished breakfast before the
man at the masthead shouted:

"Sail ho!" The captain rushed upon deck,
exclaiming:

"Masthead, there!"

"Ay, ay, sir."

"Where away is the sail?" The precise answer to this question I do not recollect but the
captain proceeded to ask:

"What does she look like?"

"A square rigged vessel, sir," was the reply of
the lookout. A few minutes afterward, the captain shouted again:

"Masthead, there!"

"Ay, ay, sir."

"What does she look like?"

"A large ship, sir, standing toward us."

By this time most of the crew were on deck,
eagerly straining their eyes to obtain a glimpse
of the approaching ship and murmering their
opinions to each other on her probable character.
Then came the voice of the captain shouting:

"Keep silence, fore and aft!" Silence being
secured, he hailed the lookout:

"What does she look like?" to which the
lookout replied:

"A large frigate, bearing down upon us, sir."

A whisper ran along the crew that the stranger
was a Yankee frigate. The thought was con-
firmed by the command:

"All hands clear ship for action, ahoy!"

The drum and fife beat to quarters, bulkheads
were knocked away, the guns were released from
their lashings and the whole dread paraphernalia
of battle was produced. And, after the lapse of a
few minutes of hurry and confusion, every man
and boy was at his post, ready to do his best
service for his country—except the band of musi-

THE UNITED STATES RAKING THE MACEDONIAN.
By courtesy of Century Magazine.

cians who came aft in a body and claimed exemption from the affray by virtue of their contract. And, with the Captain's permission, they safely stowed themselves away in the cable tier.

We had only one sick man on the list [the boatswain] and he, at the cry of battle, hurried from his cot, feeble as he was, to take his post of danger. A few of the junior midshipmen were stationed on the berthdeck below with orders given in our hearing to shoot any man who attempted to run from his quarters.

Our men were all in good spirits, though they did not scruple to express the wish that the coming foe was a Frenchman rather than a Yankee. We had been told by the Americans on board, that frigates in the American service carried more and heavier metal than ours. This, together with our consciousness of superiority over the French at sea, led us to a preference for a French antagonist.

The Americans among our number felt quite disconcerted by the necessity that compelled them to fight against their own countrymen. One of them, named John Card, as brave a seaman as ever trod a plank, ventured to present himself to the captain as a prisoner, frankly declaring his objections to fight. Captain Carden, very ungenerously ordered him to his quarters, threatening

to shoot him if he made the request again. Poor
fellow! He obeyed the unjust command and was
killed by a shot from his own countrymen.

As the approaching ship showed American
colors, all doubt of her character was at an end.
" We must fight her " was the conviction in every
breast. Every possible arrangement that could
insure success was accordingly made. The guns
were shotted, the matches [long pieces of slow-
burning rope for igniting guns] were lighted;
for, although our guns were furnished with
first-rate locks, they were also provided with
matches, attached by lanyards, in case the lock
should miss fire.

A lieutenant then passed through the ship
directing the marines and boarders, who were
furnished with pikes, cutlasses and pistols, how
to proceed if it became necessary to board the
enemy. He was followed by the captain who
exhorted the men to fidelity and courage, urging
upon their consideration the well-known motto
of brave Nelson: " England expects every man
to do his duty."

In addition to all these preparations on deck,
some men were stationed in the tops with small
arms, whose duty it was to attend to trimming
the sails and to use their muskets, provided we
came to close action. There were others also be-

low, on deck, called "sail trimmers" to assist in working the ship, should it be necessary to shift her position during the action.

My station was at the fifth gun on the main deck. It was my duty to supply my gun with powder, a boy being appointed to each gun in a ship on the side we engaged, for that purpose. A woolen screen, saturated with water, was placed before the entrance to the magazine, with a hole in it, through which cartridges were passed to the boys. We received them there and, covering them with our jackets [to prevent sparks from prematurely exploding them] hurried to our respective guns. These precautions are taken to prevent the powder taking fire before it reaches the gun.

Thus we all stood, awaiting orders, in motionless suspense. At last we fired three guns from the larboard [port] side. This was followed by the command:

"Cease firing; you are throwing away your shot." Then came the order:

"Wear ship and prepare to attack enemy with your starboard guns."

Soon after this I heard firing from some other quarter which I, at first, supposed to be a discharge from our quarter deck guns, though it proved to be the roar of the enemy's cannon. A strange noise, such as I had never before heard, next arrested my attention. It sounded like the tearing of sails, just over our heads. This, I soon ascertained, was the wind or whistling of the enemy's shot through the air.

After a few minutes cessation, the firing recommenced. The roaring of cannon could now be heard from all parts of our trembling ship and, mingling as it did with that of our foe, it made a most hideous noise. By and by I heard the shot strike the side of our ship.

Then the whole scene grew indescribably confused and horrible. It was like some awfully tremendous thunder storm, whose deafening roar is attended by incessant streaks of lightning, carrying death in every flash and strewing the ground with victims of its wrath. Only, in our case the scene was rendered more horrible than that by the presence of torrents of blood which dyed our decks.

Though the recital may be painful yet, as it will reveal the horrors of war and show at what a fearful price a victory is won or lost, I will record the incidents as they met my eye during the progress of this dreadful fight.

I was busily supplying my gun with powder when I saw blood suddenly fly from the arm of a man stationed at our gun. I saw nothing strike him. The effect alone was visible. In an instant, the third lieutenant tied his handkerchief round the wounded arm and sent the groaning wretch below to the surgeon.

The cries of the wounded now rang through all parts of the ship. These were carried to the cockpit as fast as they fell, while those more fortunate men who were killed outright were immediately thrown overboard [to that shark!]

As I was stationed but a short distance from the main hatchway, I could catch a glimpse at all who were carried below. A glance was all I could indulge in for the boys belonging to the guns next to mine were wounded in the early part of the action and I had to spring with all my might to keep three or four guns supplied with cartridges.

I saw two or three of these lads fall nearly together. One of them was struck in the leg by a large shot. He had to suffer amputation above the wound. The other had a grape or canister shot sent through his ankle. A stout Yorkshire man lifted him in his arms and hurried him to the cockpit. He had his foot cut off and was thus made lame for life.

Two of the boys stationed on the quarterdeck were killed. They were both Portuguese. A man who saw one of them killed, afterward told me that his powder [cartridge] caught fire and burnt the flesh almost off his face. In this pitiable condition the agonized boy lifted up both hands, as if imploring relief, when a passing shot cut him in two.

I was an eye-witness to a sight equally revolting. A man named Aldrich had one of his hands cut off by a shot and, almost at the same moment, he received another shot which tore open his bowels in a terrible manner. As he fell, two or three men caught him in their arms and (as he could not possibly live) threw him overboard to find relief by death in the waves.

One of the officers in my division, also fell in my sight. He was a noble-hearted fellow named Nan Kivell. A grape or canister shot struck him near the heart. Exclaiming:

"Oh! My God!" he fell and was carried below, where, shortly afterward, he died.

Mr. Hope, our first lieutenant, was wounded by a grummet or small iron ring, probably torn from a hammock clew by a shot. He went below, shouting to the men to fight on. Having had his wound dressed, he came up again, shouting at the top of his voice, bidding us to fight with all our might.

The battle went on. Our men kept cheering

with all their strength. I cheered with them, though I confess I scarcely knew what for. Certainly there was nothing very inspiriting in the aspect of things where I was stationed. So terrible had been the work of destruction around us, it was termed a slaughter house. Not only had we had several boys killed or wounded but several of the guns were disabled. The one I belonged to had a piece of the muzzle knocked out and when the ship rolled, it struck a beam of the upper deck with such force as to become jammed and fixed in that position.

A 24-pound shot also passed through the screen of the magazine, immediately over the orifice through which we passed the powder. The school master received a death wound. The brave boatswain, who came from the Sick Bay to the din of battle, was fastening a stopper to a back stay which had been shot away, when his head was smashed to pieces by a cannon ball. Another man, going to complete the unfinished task, also was struck down. Another of our midshipmen received a severe wound.

The unfortunate wardroom steward, who attempted to cut his throat on a former occasion, was killed. A fellow named John, who for some petty offense had been sent on board as punishment, was carried past me wounded. I distinctly heard the large blood-drops fall pat, pat, pat on the deck. His wounds were mortal. Even a poor goat, kept by the officers for her milk, did not escape the general carnage. Her hind legs were shot off and poor Nan was thrown overboard.

CHAPTER X.

SCENES AFTER THE BATTLE.

(*Leech's narrative concluded.*)

Such were the terrible scenes amid which we
kept on our shouting and firing. Our men fought
like tigers. Some of them pulled off their jackets,
others their jackets and waists; while some, still
more determined, had taken off their shirts and
with nothing but a handkerchief tied around the
waistbands of their trousers, fought like heroes.
Jack Sadler was one of these.

I also observed a boy named Cooper stationed
at a gun some distance from the magazine. He
came to and fro on the full run and appeared to
be as "merry as a cricket." The third lieutenant
cheered him along occasionally by saying:

"Well done, my boy. You are worth your
weight in gold."

I have often been asked what my feelings were
during this fight. I felt pretty much as I sup-
pose every one does at such a time. That men
are without thought when they stand amid the
dying and dead is too absurd an idea to be entertained for a moment. We all appeared cheerful

but I know that many a serious thought ran through my mind. Still, what could we do but keep up a semblance at least of animation?

To run from our quarters would have been certain death at the hands of our own officers. To give way to gloom and fear or to show fear would do no good and might brand us with the name of cowards and insure certain defeat. Our only true philosophy, therefore, was to make the best of our situation by fighting bravely and cheerfully. I thought a great deal, however, of the other world. Every groan, every falling man told me that the next instant I might be before the Judge of all the earth.

For this, I felt unprepared. But, being without any particular knowledge of religious truth, I satisfied myself by repeating again and again the Lord's prayer and promising that, if spared, I would be more attentive to religious duties than ever before. This promise, at the time, I had no doubt of keeping. But I have since learned that it is easier to make promises, amidst the roar of cannon in battle or in the horrors of shipwreck, than to keep them when danger is over and safety smiles upon our path.

While these thoughts secretly agitated my mind, the din of battle continued. Grape and canister shot were pouring through our port holes like leaden rain, carrying death in their train. The large shot crashed against the sides of the ship like monstrous sledge hammers, shaking her to the very keel, or passing through her timbers scattered terrific splinters which did more appalling work than even their death-dealing blows.

Some idea may be formed of the effect of grape and canister when it is known that grapeshot is formed by seven or eight balls confined to an iron ring and tied in a cloth. These balls are scattered by the explosion of the powder. Canister shot is made by filling a powder canister with balls, each as large as two or three musket balls. These also scatter with direful effect when discharged.

What, then, with splinters, cannon balls, grape and canister poured incessantly upon us, you may be assured that the work of destruction went on in a manner which must have been satisfactory even to the King of Terrors himself.

Suddenly, the rattling of the iron hail ceased. We were ordered to stop firing. A profound silence ensued, broken only by the stifled groans of the brave sufferers below. It was soon ascertained that the enemy had forged ahead to repair damages, for she was not so disabled but what she could sail without difficulty while we were so utterly cut up that we were completely helpless.

8

Our head braces were shot away. The fore and main topmasts were gone. The mizzen mast hung over the stern, having carried several men over with its fall. We were in a state of complete wreck.

A council was now held among the officers on the quarterdeck. Our condition was perilous in the extreme. Victory or escape were alike impossible. Our ship was disabled, many of our men were killed and many more wounded. The enemy would, without doubt, bear down on us in a few moments and, as she could now choose her own position, would rake us fore and aft. Any further resistance, therefore, was folly. So, in spite of Lieutenant Hope, who was for fighting to the last and sinking alongside, it was determined to strike our bunting.

This was done by the hands of a brave fellow named Watson, whose saddened brow told how severely it pained his lion heart to do it. To me it was a pleasing sight, for I had seen enough fighting for one Sabbath. His Britannic Majesty's frigate *Macedonian* was now a prize of the American frigate *United States*.

[When the news of this battle reached England Lord Churchill, very kindly, sent a copy of Captain Carden's official report with the list of killed and wounded to Samuel's mother at Blen-

heim ; calling attention to the fact that Leech's
name did not appear in the list of casualties.]

Immediately upon the surrender I went below
to see how matters appeared there. The first
object I met was a man bearing a limb which had
just been detached from some suffering wretch.
The leg was thrown overboard. Pursuing my
way to the wardroom I necessarily passed
through the steerage, which was strewn with the
wounded.

It was a sad spectacle, made more appalling by
the groans and cries which rent the air. Some
were groaning, others were swearing most bit-
terly, a few were praying, while those last arrived
were begging most piteously to have their wounds
dressed next. The surgeon and his mate were
smeared with blood from head to foot. They
looked more like butchers than doctors.

Having so many patients, they had once shifted
their quarters from the cockpit to the steerage.
They now moved to the wardroom and the long
table, round which the officers had sat over many
a merry feast, was soon covered with bleeding
forms of maimed and mutilated seamen.

While looking around the wardroom I heard
a noise above, occasioned by the arrival of the
boats from the conquering frigate. Very soon a
lieutenant, I think his name was [John B.] Nich-

olson, came into the wardroom and said to the busy surgeon:

"How do you do, doctor?"

"I have enough to do," replied he shaking his head sadly, "You have made wretched work for us."

These officers were not strangers to each other for they had met when the two frigates were lying at Norfolk, some months before.

I now set to work to render all the aid in my power to the sufferers. Our carpenter, named Reed, had his leg cut off. I helped to carry him to the after wardroom; but he soon breathed out his life there—and then I assisted in throwing his mangled remains overboard.

We got out the cots as fast as possible for most of the wounded were stretched out on the gory deck. One poor fellow who laid with a broken thigh, begged me to give him a drink of water. I gave him some. He looked unutterable gratitude, drank—and died.

It was with exceeding difficulty that I moved through the steerage, it was so covered with mangled men and so slippery with streams of blood. There was a poor boy crying as if his heart would break. He had been the servant to the bold boatswain whose head had been dashed to pieces by a cannon ball. Poor boy! He felt that he lost his only friend. I tried to comfort him by reminding him that he ought to be thankful for having escaped death himself.

Here also I met one of my messmates who showed the utmost joy at seeing me alive for he had heard that I had been killed. He was looking up his messmates which, he said, was always done by sailors after a battle. We found two of our mess wounded. One was the Swede, Logholm, who came so near drowning, a few months before. We held him while the surgeon cut off his leg above the knee.

The task was most painful to behold, the surgeon using his knife and saw on human flesh and bones as freely as the butcher at the shambles on the carcass of a beast. Our other messmate suffered still more than the Swede. He was sadly mutilated about the legs and thighs with splinters. Such scenes of suffering as I saw in that wardroom I hope never to witness again.

Most of our officers and men were taken on board the victor ship. I was left, with a few others, to take care of the wounded. My master, the sailing master, was also among the officers who continued in our ship. Most of the men who remained were unfit for any service, having broken into the spirit room and made themselves drunk. Some broke into the purser's room and helped themselves to clothing while others, by previous agreement, took possession of their dead messmates' property.

For my own part I was content to help myself to a little of the officers' provisions which did me more good than could be obtained from rum. What was worse than all, however, was the folly of the sailors in giving spirits to their wounded messmates, since it only served to aggravate their distress.

Among the wounded was a brave fellow named Wells. After the surgeon had amputated and dressed his arm, he walked about in fine spirits— as if he had received only a slight injury. Indeed, under the operation he manifested a similar heroism, observing to the surgeon:

"I have lost my arm in the service of my country but I don't mind it doctor, it's the fortune of war."

Cheerful and gay as he was he soon died. His companions gave him rum; he was attacked by fever and died. Thus his messmates actually killed him with kindness.

We had all sorts of dispositions and temperaments among our crew. To me it was a matter of great interest to watch their various manifestations. Some who had lost their messmates, appeared to care nothing about it while others were grieving with all the tenderness of women. Of these was the survivor of two seamen who had formerly been soldiers in the same regiment. He bemoaned the loss of his comrade with expressions of profoundest grief.

There were, also, two boatswain's mates named Adams and Brown, who had been messmates for several years in the same ship. Brown was killed or so wounded that he died soon after the battle. It was really a touching spectacle to see the rough, hardy features of the brave old sailor, streaming with tears as he picked the dead body of his friend from among the wounded and gently carried it to the ship's side, saying to the inanimate form he bore:

" Oh Bill! We have sailed together in a number of ships, we have been in many gales and some battles. But this is the worst day I have seen! We must now part! "

Here he dropped the body into the sea and then a fresh torrent of tears streaming over his weather-beaten face, he added:

" I can do no more for you, Bill. Farewell! God be with you! "

Here was an instance of genuine friendship, worth more than the heartless professions of thousands who, in the fancied superiorty of their elevated position in the social circle, will deign nothing but a silly sneer at this record of a sailor's grief.

It was a rather singular circumstance that, in both the contending frigates, the second boatswain's mate bore the name of William Brown; and that they were both killed. Yet such was the fact.

The great number of wounded kept our surgeon and his mate busily employed at their horrid work until late at night and it was a long time before they had much leisure.

I remember passing round the ship on the day after the battle. Coming to a hammock, I found some one in it, apparently asleep. I spoke. He made no answer. I looked into the hammock— he was dead. My messmates coming up, we threw the corpse overboard—it was no time for useless ceremony. The man had probably crawled into his hammock (badly wounded) the day before and, not being noticed in the general distress, bled to death!

When the prize crew from the *United States* first boarded our frigate to take possession our men, heated with the fury of the battle, exasperated by the sight of their dead and wounded shipmates (and rendered dangerous by the rum they had obtained from the spirit room) felt and exhibited some disposition to fight their captors.

But after the confusion had subsided and part of our men were snugly stowed away in the American ship and the remainder found themselves kindly used in their own, the utmost good feeling began to prevail.

We took hold and cleansed the ship, using hot vinegar to take out the scent of the blood that dyed the white of our planks with crimson. We also took hold and aided in fitting out our disabled frigate for her voyage. This [after two days of hard work] being accomplished, both ships sailed in company for the American coast.

I soon felt myself perfectly at home with the American seamen; so much so that I preferred to mess with them. My shipmates also participated in similar feelings in both ships. All idea that we had been trying to shoot each other's brains out so shortly before, seemed forgotten. We ate together, drank together, joked, sang, laughed and told yarns. In short, a perfect union of ideas, feelings and purposes seemed to exist among all hands.

A corresponding state of unanimity existed, I was told, among the officers. Commodore Decatur showed himself to be a gentleman as well as a hero in his treatment of the officers of the *Macedonian*. When Captain Carden offered his sword to the commodore, he remarked as he did so:

"I am an undone man. I am the first British naval officer that has struck his flag to an American!"

The noble commodore either refused to receive the sword or immediately returned it, smiling as he said:

"You are mistaken, sir. Your *Guerrière* has been taken by us, so the flag of a frigate was struck before yours."

This news somewhat revived the spirits of the old captain. But, no doubt, he still felt his soul sting with shame and mortification at the loss of his ship. Participating as he did in the haughty spirit of the British aristocracy, it was natural for him to feel galled and wounded to the quick in the position of a conquered man.

We were now making the best of our way to America. Notwithstanding the patched up condition of the *Macedonian,* she was far superior in a sailing capacity to her conqueror. The *United States* had always been a dull sailer and had been christened by the name of *Old Wagon.* Whenever a boat came alongside of our frigate and the boatswain's mate was ordered to "pipe away" the boat's crew, he used to sound his shrill call on the whistle and bawl out:

"Away, *Wagoners,* away!" instead of "Away, *United States* men, away!"

This piece of pleasantry used to be rebuked by the officers but in a manner that showed that they enjoyed the joke. They usually replied:

" Boatswain's mate, you rascal, pipe away *United States* men, not *Wagoners.* We have no wagoners on board of a ship! "

Still, in spite of rebuke, the joke went on until it grew stale by repetition. One thing was made certain, however, by the sailing qualities of the *Macedonian;* which was, that if we had been disposed to escape from our foe before the action, we could have done so with all imaginable ease. This, however, would justly have exposed us to disgrace, while our capture did not.

(*End of Leech's narrative.*)

CHAPTER XI.

TRIUMPHANT RETURN TO PORT.

Captain Carden had entered into this momentous battle with consummate skill and shrewdness—but he was fatally mistaken in his surmises. He at first supposed that his antagonist was the American frigate *Essex* which, as he knew, was armed almost entirely with short range guns and, quickly discovering that he had the superiority in sailing, he held his ship at long range so that his long 18-pounders would be effective beyond the reach of the American short cannon. He soon discovered his mistake and gallantly came to close quarters—but not before he had sustained irreparable damage.

While the loss in the *Macedonian* had been frightful (amounting to more than one-third of the entire ship's company), as so graphically described by young Leech, the casualties in the *United States* were insignificant: there being only five men killed and seven wounded as opposed to the thirty-six killed and sixty-eight wounded in the English ship. The *United States* was only ten or twelve feet longer than her rival and,

though being inferior in sailing qualities, she
was stronger and better armed and equipped—as
Captain Carden well knew in advance. The following table will fairly represent the strength of
the contending frigates:

COMPARATIVE FORCES AND LOSSES.

	Guns	Lbs.	Crew	Killed	Wounded	Total	Time
United States	54	787	478	5	7	12	1h. 30m.
Macedonian	49	555	297	36	68	104	

Scarcely any injury was sustained in the
American's hull or rigging so that, after splicing
some ropes and making a few repairs, she was
in a condition to enter upon a similar action two
hours after the *Macedonian* surrendered. The
English frigate, on the other hand, was completely dismantled—clearly showing the immense
superiority of American naval construction, armament and equipment.

When the news of this second frigate action
reached England, it created the deepest gloom.
At first it was not believed for the London Times,
in its issue of December 28, 1812, said: "There
is a report that another English frigate, the
Macedonian, has been captured by an American.
We shall certainly be very backward in believing
a second recurrence of such a national disgrace.
. . . . We have heard that the statement is dis-
credited at the Admiralty but we know not on

what precise grounds. Certainly there was a
time when it would not have been believed that
the American navy could have appeared upon
the seas after a six months' war with England;
much less that it could, within that period, have
been twice victorious. *Sed tempora mutantur!*"

The uncomfortable suspicion evidently grew
and, on the next day, the acute pang of confirma-
tion extorted from the British lion the following
cry: "Oh miserable advocates! Why, this
renders the charge of mismanagement far heavier
than before! In the name of God, what was
done with this immense superiority of force?
Why was not a squadron of observation off every
port which contained an American ship of war?
Why was not Rodgers intercepted with his whole
squadron and taken within sight of his own
coasts?"

On the following day The Thunderer's rage
subsided into the following lament: "Oh, what
a charm is hereby dissolved! What hopes will
be excited in the breasts of our enemies! The
land spell of the French is broken [alluding to
Napoleon's disastrous retreat from Moscow] and
so is our sea spell."

In its issue of December 26, 1812, The London
Morning Chronicle, with greater moderation
asks: "Is it not sickening to see that no ex-

perience has been sufficient to rouse our Admiralty to take measures that may protect the British flag from such disgrace?"

Although the *United States* had won a signal victory over the *Macedonian*, it was yet far from certain that she would secure the fruits of the battle. The vessels were on the eastern side of the Atlantic and had a long distance to sail before they could reach an American port.

Young Leech describes the home passage as follows: "Our voyage was one of considerable excitement. The seas swarmed with British cruisers and it was extremely doubtful whether the *United States* would elude their grasp and reach the protection of an American port with her prize. I hoped most sincerely to avoid them, as did most of my old shipmates. Our former officers, of course, were anxious for a sight of a British flag.

"But we saw none and, after a prosperous voyage, the welcome cry of 'Land ho!' was heard."

Entering the eastern end of Long Island Sound, the frigates made the port of New London. The *United States* came safely to anchor but the *Macedonian*, owing to a sudden shift in the wind, was compelled to remain in the offing several hours. Finally, fearing that she

might fall into the hands of a hostile squadron,
she made for Newport where the prize was
warmly greeted.

By this time most of the English wounded
were well on the road to recovery. The last one
to die was Thomas Whittaker, who had been
badly injured by splinters. He suffered such
pain that he finally became crazed so that it was
necessary to confine him. Just before land was
sighted he was mercifully relieved from his
sufferings by death. Sewing up his body in his
hammock his messmates placed it on a grating in
a bow port. Midshipman Archer of the *Macedonian* read the beautiful burial service of the
Church of England and at the words:

"We commit the body of our brother to the
deep," the grating was elevated and, amid profound silence, the body splashed heavily into the
sea.

The wounded were now sent ashore where
they received every attention while the prisoners
were confined in a barn under a not very strict
guard; for it appears many of them escaped—not
to return to the British service but to keep away
from it.

After a short stay at Newport the *Macedonian*
got under weigh and, joining the *United States*
at New London, both ships proceeded to New

AN 1812 POWDER BOY CARRYING CARTRIDGES
Drawn from a contemporaneous sketch.

York by the Hell Gate route; the men in both ships being kept busy answering cheers from passing craft.

Dropping anchor near Ward Island, the frigates were visited by many thousand people and as Samuel records: "Finding them extremely inquisitive and being tolerably good natured myself, I found profitable business in conducting them about the ship, describing the action and pointing out the places where particular individuals fell. For these services I gained some money and much good will.

"The people who had been to see us, on returning to the shore, used to tell how an English boy had conducted them all over the ship and told them the particulars of the fight. It soon became quite common, for those who came afterward, to inquire if I was 'that English boy taken' in her?"

It was by means of the money and good will thus earned that Samuel finally made his escape from the British service. Of course, the American officers could not let him go free because they were responsible for their prisoners so that, when an exchange was effected, they could be produced. Then again, as Leech well knew, if he attempted to escape a horrible punishment awaited him from the lash or noose—possibly

9

both—if he ever again was caught by the British, for his act would be deemed desertion. Furthermore, the Admiralty offered special bounty for the apprehension of any deserter who had been captured by the Americans.

Fully alive to the danger attending it, Samuel determined to escape and made his plans accordingly. Mr. Tinker, the pilot who took the frigates from New London to New York, very kindly offered to take Leech as an apprentice if he once got free; in fact, many of the visitors to the frigate were so pleased with the "bright English lad" that they offered him any assistance he might desire.

No time was to be lost, however, if Leech was to embrace this, probably, his only opportunity ever to escape from the British for he learned that a cartel was on its way to New York for the purpose of conveying all the English officers and seamen to Halifax. Indeed, the boat was expected to arrive at any moment.

On the day before Christmas there was an unusual number of visitors aboard the prize frigate and, as there was a large proportion of women who could not very well climb the rope ladder up the *Macedonian's* steep sides, Captain Carden very gallantly caused a hogshead to be rigged so that after one head had been knocked in and part of the front cut out, a comfortable " elevator car " was produced by placing a seat athwart it.

This " hogshead " was lowered from a yard-arm into the shore boat alongside of the frigate. One lady at a time seated herself in the " car " and, throwing a flag around her feet, was merrily hoisted up by a gang of jolly tars. Gaining the level of the deck, the hogshead was swung inboard and the lady could step out with ease. " This contrivance," records Leech, " afforded a great deal of amusement and kept the British officers and merry tars agreeably busy waiting on their fair visitors.

It was when this scene of unusual activity was at its height, that Samuel made his bold dash for liberty. He made arrangements with the American boatswain, Mr. Dawson, to have his clothing sent to New York if he (Leech) succeeded in getting clear. Noticing a small colored boy in a boat alongside the *Macedonian*, Leech quietly asked :

" Can you tell me where I can get some geese and turkeys on shore for our officers? "

" I guess you can get some at the houses," responded the youth."

" Well, then," continued Leech, " will you set me ashore. I want to get some for our officers." To this the colored boy replied :

"Yes, if you will go and ask my master who is on board your ship."

This was an obstacle to Samuel's plans he had not foreseen. He knew that the master would not give the desired permission so he gave over the attempt in despair.

Going below, he met one of his shipmates, a boy two years younger than himself, named James Day. Leech revealed to him his plan to escape and urged him to go with him. Day at first declined to take the risk, giving as his reason that he had no money with which to pay expenses.

"But I have money," replied Leech, "and as long as I have a shilling you shall have half of it."

"But I am afraid we cannot get away without being caught and so get a thorough flogging—and perhaps be hung," protested Day.

A new idea had now struck our hero and clutching Day by the arm he said:

"Never mind that. I have contrived that business. The boat's waiting to set us on shore. Come along, Jim, don't be frightened. 'Nothing venture, nothing have,' you know. Come, come. Here's the boat alongside," and, fairly dragging the boy, Leech returned to the gangway and boldly assured the colored lad that his master had given the desired consent, provided haste was made. The two runaways jumped into the boat and, in a jiffy, were being pulled toward the shore.

This was the critical moment of the undertaking and Samuel's heart was fairly pumping with anxiety and fear. Every moment he imagined he heard a stern command from the frigate to return. Every rattle aboard he construed to be a pursuing boat making after him.

At one moment his heart fairly jumped into his throat, when a harsh voice did hail them from the frigate. It was not from a British officer, however, but from the colored boy's master who shouted out:

"Where are you going with that boat?"

Recovering from the fright, Leech pursuaded the negro that his master was only bidding him to make haste, so the lad replied:

"I am going to get some geese, sir," and pulled on so they were beyond reach of hearing. In another moment Leech, for the first time and to his unspeakable delight, stood on American soil— a free lad. He never learned if the colored boy's master ever asked if he was not the biggest "goose" he was going after. Leech gladly gave him half a dollar and set out on foot for New York, some ten miles distant.

A ten-mile walk, in former days was a mere trifle to the sturdy English runaway. Many a time had he covered that distance among the pleasant fields and parks about Wanstead and Blenheim. But now he discovered that he had been aboard ship so long that less than half that distance exhausted him—it required time to regain his "land legs." So, when yet some distance from the city, the boys put up at a roadside tavern.

The inmates seemed somewhat surprised to see two lads asking for lodgings in such a confident manner but, when they learned that they were deserters from the British frigate, they extended every hospitality. Forming a circle round the fireplace in the public room, they listened attentively to the narrative and songs of the runaways. At a late hour the boys were shown into a clean room and, for the first time in years, they slept in a bed.

"It seemed strange to us," recorded Leech, "to find ourselves in a bed after sleeping so long in hammocks. Nevertheless we slept soundly and, to our inexpressible pleasure, arose on the following morning at our leisure instead of being driven out by a swearing boatswain at our heels."

After a hearty breakfast (doubly delicious on account of the shore cooking and fresh provisions) Leech, with all the dignity he could summon, marched proudly up to the "captain's office " and asked for a settlement, clinking the money noisily in the palm of his hand as an earnest of his good faith and " financial ability." Truly, it was one of the proudest moments of his life. He was somewhat crestfallen (though none the less pleased) when the host refused to take a cent for the entertainment.

Three days after Leech's escape from the *Macedonian,* the cartel arrived and on the same day sailed for Nova Scotia with the remaining English prisoners. Lucky, indeed, was it that our hero made the venture when he did.

CHAPTER XII.

TRIALS OF A DESERTER.

Arriving in New York the runaways met several other deserters from the *Macedonian* and through them found lodgings in a sailors' boarding house kept by a widow named Elms, near the old Fly Market in Front Street. After spending a week in gratifying a natural desire to see a strange city, Leech was startled one day by the roaring of cannon. It proved to be salutes in honor of the *United States'* and *Macedonian's* arrival in the Brooklyn Navy Yard, those ships having made the, then, dangerous passage through Hell Gate.

Leech now ventured aboard his old home ship, to get his clothes from Boatswain Dawson as pre-arranged—and the first person he met gave him a fright. It was none other than Lieutenant Nicholson of the *United States* who eyed our hero sharply but afterward gave him a kindly reception. The American sailors, also congratulated him on his success in getting clear of the British frigate.

It was about this time that the citizens of New

York gave to the officers and men of the *United States* a public dinner at the City Hotel and free admission to a theatrical entertainment. As a promising citizen of America, young Leech was invited to attend. He accepted but, on overhauling his wardrobe, found that his English uniform would be unpleasantly conspicuous on account of the regulation buttons. This difficulty was overcome, however, by the skill of his widowed landlady who managed to cover the metal buttons with blue cloth.

This celebrated dinner (and after performance) is a matter of history but Leech has thrown some interesting side-lights on it in his own peculiar fashion. He said the dinner was " followed by more than the usual amount of drinking, laughing and talking; for, as liquor was furnished in great abundance, the men could not resist the temptation to get drunk. As they left the room to go to the theater, the poor plates on the sideboard proclaimed that ' Jack was full three sheets in the wind.' Almost everyone, as he passed, gave them a crack crying out as they fell:

" ' Save the pieces.' "

At the theater Samuel saw Decatur and records: " I was much struck with the appearance of Decatur that evening as he sat in full uniform, his pleasant face alive with the excitement of the occasion. He formed a striking contrast to the appearance he made when he visited our ship on the passage to New York. Then he wore an old straw hat and a plain suit of clothes which made him look more like a farmer than a naval commander."

Leech concludes his account by saying that, after the theater, the men were ordered to report to the frigate the next morning but:

" It was a week before they all returned."

Another interesting piece of information Samuel gives us relates to a little trick these Yankee tars played on Decatur. He says:

" Of course, this profusion of praise turned the brains of some of these old tars and at every opportunity they would steal ashore for a spree. This brought them into trouble and some to the gangway to be flogged. To avoid the punishment the foxy old salts would visit the commodore's lady with some piteous tale, begging her to intercede for them with the captain. This she did with almost constant success. The lucky tar would then go on board telling his shipmates that:

" ' She has a soul to be saved.' "

Mention has been made in these pages of Jack Sadler, the bosom friend of the redoubtable Bob Hammond. Sadler managed to get ashore and became an enthusiastic Yankee. He enlisted in the army and was quartered at Hartford. One Sunday his company was marched to church and the good minister announced as his text:

"Fear God and honor the King." Jack, with vivid recollection of many a cruel lash on his back, so far forgot himself as to jump up in his pew and shout: "Don't let us hear about the king—but about Congress!"

The dangers and difficulties English seamen experienced in entering the American service in those days, is amusingly described by Leech as follows: "One day, I was sauntering around the wharves with my companion, James Day, when I met a number of the *Macedonian* crew who had shipped aboard the *John Adams* and they dragged me aboard with them.

"To avoid being detected it was usual for our men to assume new names and to hail from some American port. I had some objection to this, as I feared it might bring me into the awkward dilemma of the Irishman who was caught aboard an American vessel by a British cruiser. After he declared himself an American the officer asked him:

"'What part of America did you come from?'

"'I used to belong to Philmadelph but now I belong to Philmaph York,' replied Paddy, in a vain endeavor to conceal the 'flannel' in his brogue.

"'Well, can you say peas?' continued the officer.

"'Pase, sir,' answered Pat and he was duly transferred to the English ship."

On the advice of his former shipmates, Leech assumed the name of William Harper from Pine Street, Philadelphia; and, going aboard the corvet, was duly paraded before the officers for "inspection." One of them said:

"Well, my boy, what is your name?"

"William Harper, sir," confidently responded the lad.

"What part of America do you belong to?"

"Philadelphia, sir." Here one of the officers smiled and remarked: "Ah, a fellow townsman. What street in Philadelphia?"

"Pine street, sir," replied Samuel with the expression of one who was being drawn into a net.

"What street joins Pine street, my lad?" continued his tormenter with a knowing laugh.

"I don't remember, sir," said Leech with fast ebbing confidence.

"Ah, you don't remember, do you? Quite possibly," said the mischief-loving officer, for he knew pretty well all about Samuel's antecedents. "But, at least, you can tell us in what state Philadelphia is situated?"

This was a poser for the poor lad and, thinking

to get off with the honors of war he gaspingly replied (not really knowing what he was saying) :

" Gentlemen, it is so long since I have been in Philadelphia that—that—I—I—I—really forget what state the city is in unless it is in the state of rest."

This answer seemed to please some of the officers immensely, for they burst out laughing : but the one questioning Leech (he who claimed to be " a fellow townsman ") appeared vexed, for he pointed to one of Leech's English buttons, which had (unknown to him) relieved itself of its cloth covering, and said:

" Where did you get that English button ? Did you pick it up in Philadelphia ? "

This was a shot that raked Samuel fore and aft. He hauled down his colors and remained silent. The officers laughed heartily and one of them said:

" Go below, my lad ; you will make a pretty good Yankee."

The next morning Leech was taken ashore to sign the shipping papers but, with that strong commonsense, characteristic of him, he argued to himself that there were too many men from the *Macedonian* already aboard the *John Adams* and, if she were captured there would be small chance of escaping discovery and a noose at the yardarm would be his reward.

So he did not enlist in her. This decision was backed by his knowledge that strict orders had been issued from the Admiralty to keep a sharp watch for men who had been captured by American war ships. While Samuel's reasoning was correct in theory, the fact was that the *John Adams* was not captured by the enemy.

After two weeks of idleness and, finding that his stock of money accumulated aboard the *Macedonian* was fast ebbing away, Leech accepted the offer of an Englishman by the name of Smith (who was a deserter from the British army but was then employed as a bootmaker in the firm of Benton & Co., in Broadway) to become an apprentice in the "art, science, secrets and mysteries of a cordwainer."

" Behold me, then," records our hero, " transformed from the character of a runaway British sailor into that of a quiet scholar, at the feet of St. Crispin ; where, in the matter of awls, waxends, lapstones and pegs, I soon became quite proficient."

It is altogether likely that our hero would have passed the remainder of his days in the " art " of shoemaking had it not been for a rumor that reached him one day, after he had been about two months in his new service, that a tall, stout sailor named George Turner was in the crew of the

AN AMERICAN MAN-O'-WARSMAN IN 1812.
Drawn from a contemporaneous sketch.

United States. Determined to investigate, Samuel, one fine Sunday morning went aboard that frigate and was heartily received by some of his former shipmates of the *Macedonian* who had entered the American service.

Leech soon presented himself to his cousin and, after reminding him of several incidents connected with their relatives and home at Wanstead, established his identity before that worthy tar. Turner advised the youth to give up the sea and, very kindly, offered him a home in Salem where he had a wife and family.

Severing his connection with his kind employer in the cordwainer shop, Leech engaged steerage passage in a sloop bound for Providence, for five dollars. Before that vessel sailed, Samuel found that his clothes bag had been robbed by a negro and the master of the craft, to save the good name of his packet, returned to Leech two dollars of the passage money as compensation. From Providence he proceeded to Boston in a coach chartered for the exclusive use of a party of merry sailors.

On his arrival in Salem, Samuel was warmly welcomed by Mrs. Turner who, being very superstitious, declared that she knew of his coming because of some peculiar antics of tea leaves in her cup that morning.

A few weeks afterward the good woman
aroused Samuel early one morning and hurried
him off to the post office because she had dreamed
of "catching fish." Surely enough, our hero
soon returned with a letter containing a hundred
dollar bill from her husband.

Having no steady employment, young Leech
spent most of his time around the wharves and
shipping where he saw a number of privateers.
He also went on a number of fishing trips in
schooners. On one of these occasions, Leech and
his party came near being captured by an English
war ship. The party had been out all night and
toward morning, being tired out, nearly every one
aboard went to sleep.

Luckily, one of the party, Lewis Deal, who had
once been a quartermaster aboard the *United
States,* kept a weather eye open; for he knew
that the coast was alive with British cruisers.
Just at dawn, the report of a cannon close by,
startled every one from his slumber and Deal
exclaimed:

"There! I told you to look out for Johnny
Bull!"

Looking about, they saw an English gun-
brig in full chase of a Boston sloop, within easy
gunshot of them. Hastily weighing anchor, the
fishing party made sail and soon reached port in

safety—the gun-brig being so intent on her chase that she failed to discover the excursionists in the uncertain light of dawn. For a moment, however, Samuel has painful visions of swinging at a yard-arm which did not leave him until the brig was fairly out of gunshot.

In the summer of 1813 young Leech determined to enter the American navy and, as the *Constitution, Frolic* and *Siren* were at that time in Boston, shipping crews, he had the choice of those vessels. His preference was for *Old Ironsides* but, as his cousin Turner had once sailed under the commander of the *Siren*, George Parker, and highly commended that officer, Leech enlisted in that brig.

The *Siren*, owing to the blockade was unable to get to sea for several months. All this time her crew was exercised in various drills. Samuel records: " My first impressions of the American service were very favorable. The captain and officers were kind, while there was a total exemption from that petty tyranny exercised by the upstart midshipmen in the British service. Our men were as happy as men ever were in a man-of-war.

" We were all supplied with stout leather caps, something like those used by firemen. These were crossed by two strips of iron, covered with

10

bear skin and were designed to defend the head, in boarding an enemy's ship, from cutlass strokes. Strips of bear skin were used to fasten them on and, having the fur on, served the purpose of false whiskers and causing us to look as fierce as hungry wolves."

CHAPTER XIII.

SIREN'S LIVELY CRUISE.

Early in June, 1814, the *Siren* was ready for sea and, getting under weigh in company with the famous privateer *Grand Turk,* stood down the harbor bound for a cruise on the west coast of Africa. In passing the fort, the *Siren* received the usual hail:

"Brig, ahoy! Where are you bound to?"

To this First Lieutenant John B. Nicholson (whom Samuel had met in the *United States* after her action with the *Macedonian*) jocosely replied:

"There and back again, on a man-of-war's cruise!"

"Such a reply," said Leech, "would not have satisfied a British sentry but we shot past the fort unmolested"; the officers of the fort probably knowing, full well, the characters of the passing ships.

When two days out the *Grand Turk* parted company, not to be seen again until on the other side of the Atlantic—and then, under peculiar circumstances, as will soon appear.

Keeping a sharp lookout the *Siren* touched at the Canary Islands and then made for the coast of Africa where Captain Parker died. A service was read over his body and it was committed to the deep. Scarcely had the brig got under sail again when, to the horror of all, the *coffin was seen to be floating in the wake of the ship.* "The reason for this," said Leech, "was that the carpenter bored holes in the top and bottom, when he should have made them in the top only."

Such a grewsome accident would have deterred a more superstitious crew from continuing on the voyage but Captain Nicholson called all hands together and frankly laid the situation before them; offering the choice of returning home or of continuing the cruise. With three hearty cheers the men expressed their unanimous desire to continue and the *Siren* held a course accordingly.

Leech spoke in the highest terms of Nicholson, saying: "He was a noble-minded man, very kind and civil to his crew. Seeing me one day with rather a poor hat on, he called me aft and presented me with one of his own."

One morning the welcome cry "Sail ho!" aroused every man in the brig and attention was attracted to a strange vessel which had hove-to, with her courses hauled up. At first it was thought that she might be a British man-of-war.

The *Siren* was cleared for action and the crew sent to battle quarters but on nearer approach the stranger was recognized as their old friend, the privateer *Grand Turk*. Her commander did not seem to know the *Siren* for, after assuring himself that she was a brig of war, he crowded on all sail to escape. As Captain Nicholson did not care to chase, she was soon out of sight.

Running close along the African coast, the Americans, one day, saw several fires burning on the hills which, on investigation, they learned was the native method of indicating that they desired trade with the passing ship. The *Siren* hove-to, the negroes put off in canoes and a quantity of oranges, limes, cocoanuts, tamarinds, plantains, yams and bananas were taken aboard as welcome additions to the ship's larder. The brig remained here several days in the vain hope of falling in with English traders.

It was while here that Leech first, really, appreciated the great value of water. He records: " We began to experience the inconvenience of a hot climate. Our men were covered with blotches or boils. To make it worse, was the want of fresh water. We were placed on an allowance of two quarts a day for each man. This occasioned much suffering for, after mixing our Indian meal for pudding, our cassava [a root which, on being

ground made tolerable bread] and our whiskey
for grog, we had little left to assuage our burning
thirst.

"Some, in their distress, drank large quantities
of sea water which only increased their thirst
and made them sick. Others sought relief in
chewing lead, tea leaves or anything that would
create moisture. Never did we feel more delighted than when our boat's crew announced the
discovery of a pool of fine, clear water near the
shore. We could have joined in the most enthusiastic cold-water song ever sung."

One night, while cruising along the coast, a
large ship was discovered at anchor near the
shore. Owing to the darkness it could not be
determined whether she was a merchantman or
a man-of-war, so the utmost caution was exercised in approaching her. It was not long before
all doubt as to her character was dispelled for,
suddenly, she set sail and made chase after the
Siren.

By the aid of powerful night glasses Captain
Nicholson saw that she was a British frigate—
and "meant business." The *Siren* was cleared
for action, the cannon loaded, matches lit and the
men laid down by their guns, fully expecting to
be prisoners of war before morning; for the
wind was in a direction favorable for the frigate
outcarrying the brig.

Again visions of swinging at a yard-arm passed unpleasantly through the mind of our youthful hero for, to all appearances, it was only a question of a few hours when the steadily gaining pursuer would have the brig under her guns.

But Leech had not counted on the resourcefulness of Yankee commanders. When Nicholson realized that his powerful foe was rapidly gaining on him, he resorted to one of those tricks so successfully practiced by our privateersmen in that war.

He had purposely kept a light in full view of the frigate, as if bent on a suicidal desire to be captured. When his pursuer was nearly within gunshot, however, he rigged out a hogshead, which was sealed up and so weighted that it would float in an upright position; and on top of it he affixed a light, similar to the one he had been carrying. Dropping the hogshead carefully overboard with its decoy signal in full view, he "doused" the *Siren's* light. Then, changing his course, he made off in another direction leaving the frigate in her furious pursuit of the hogshead and its deceptive light. By daybreak the frigate was nowhere to be seen. Without doubt she "captured" the hogshead and her commander probably indulged in a prolonged soliloquy over the "singular ingenuity of these Yankees—as respects seamanship."

The next adventure the *Siren* had was equally sharp. Discovering, one night, an English merchantman at anchor in Senegal river, Captain Nicholson ran down to her and hailed. Receiving an insolent reply, he gave the order to fire—but instantly countermanded it. But it was too late. The guns had been loaded and carefully trained. The men, with burning matches, stood ready at the first order and before the countermand came they had discharged the broadside.

The swift current of the river carried the *Siren* past the merchantman, down the river. She attempted to beat up again but the unfortunate broadside had aroused the garrison of the fort, which commanded the river, and soon a rattling hail of " large size " cannon balls began to pass unpleasantly close to the heads of the Yankee crew. As it was useless to attempt the capture under such circumstances, Captain Nicholson dropped down the river, beyond the reach of the fort, to await daylight.

Next morning the merchantman was seen snugly moored under the guns of the fort and, as she was filled with soldiers and had the protection of the fort, it was clear that a stubborn resistance would be made. At first the Americans contemplated making a boat attack upon her, under cover of night. The *Siren's* crew begged

Captain Nicholson for permission but, after carefully considering the great risks he, very properly, decided to give it over.

Several men in the merchantman were, undoubtedly, killed or wounded and had it not been
for the hasty broadside, she might have been
captured by boarding and carried beyond the
reach of the fort without the garrison knowing
anything about it until morning. The *Siren's*
crew humorously dubbed this affair " The Battle
of Senegal."

After visiting Cape Three Points, Captain
Nicholson shaped his course for St. Thomas; and
it was on this run that he met his match in Yankee
nautical cunning. The English merchantman
Jane of Liverpool, was discovered and in the
hope of decoying her under his guns Captain
Nicholson displayed English colors—it not yet
being known to the Americans what the na-
tionality of the stranger was. The *Jane* promptly
responded with the Stars and Stripes and, in
return, the *Siren* showed American colors.

This was all the British master wanted to know
and, making all sail, he shaped his course for
St. Thomas which was a neutral port. The *Siren*
crowded on every stitch of canvas that would
hold the wind but the *Jane* proved to be the
better sailer of the two and gained the harbor in

safety. In the hope of catching this and another English merchantman that was in the port, the *Siren* hovered in the vicinity several days and was rewarded by a rich prize.

In a few days a sail was discovered making for St. Thomas. Hoisting English colors and dressing his officers in British uniforms (placing them in conspicuous places so that they could be readily seen by the approaching ship) Captain Nicholson—doubtlessly piqued by the trick the other Britisher had played on him and determined to show that Yankee ingenuity had not fallen below par—leisurely brought his brig within hailing distance of the Englishman when he called out:

"Ship, ahoy!"

"Hello!" was the reply.

"What ship is that?" asked the American.

"The ship *Barton.*"

"Where do you belong?"

"To Liverpool."

"What is your cargo?"

"Redwood, palm oil and ivory."

"Where are you bound to?"

"To St. Thomas."

At this moment the English flag on the *Siren* was hauled down and in its place was run up the Stars and Stripes and, to the inexpressible annoy-

ance of the Englishman, Captain Nicholson
hailed:

" Haul down your colors! "

Young Leech records: " The old captain [of
the prize] who, up to this time had been enjoying
a nap in his very comfortable cabin, now came on
deck in his shirt sleeves, rubbing his eyes and
looking so exquisitely ridiculous, it was scarcely
possible to avoid laughing. So surprised was he,
at the unexpected termination of his dreams, that
he could not command skill enough to strike his
colors; which was, accordingly, done by his mate.
As they had two or three guns aboard, and as
some of the men looked as if they would like to
fight, our captain told us, if they fired, not to
leave enough of her 'to boil a tin pot with.'
After this expressive threat, we lowered a boat
and took possession of our prize."

After taking out what goods they wanted, the
Americans set fire to her. As the flames got fully
under headway that night the burning ship presented an impressive sight which Leech has
described as follows: " It was an imposing sight
to behold the antics of the flames leaping from
rope to rope and from spar to spar until she looked
like a fire-cloud resting on the dark surface of the
water.

" Presently her spars began to fall, her masts

went by the board, her loaded guns went off, shaking up a shower of sparks which were carried high up in the heaven by the hot current of air, until they flickered out of sight. The hull was burned to the water's edge and, what was a few hours before a fine, trim ship, looking like a living creature of the deep, lay a shapeless charred mass, whose blackened outlines shadowed in the clear, still waves, seemed like the grim spirit of war seeking its prey."

The men in the *Barton* were taken to St. Thomas where they were transferred to the aforementioned *Jane*. Sailing again on a new venture, the *Siren* captured the English brig *Adventure,* laden with "monkeys, an African prince"—and other things. The monkeys were destroyed with the brig but the African who, by the way, had received a tolerably fair education in England and was strikingly polite and pleasant in his ways, shipped aboard the *Siren*. He gave his name as Samuel Quaqua.

Again returning to St. Thomas to rid herself of prisoners, the *Siren* remained in that port several days. The Americans improved this opportunity to make purchases, receiving all kinds of fruit, birds and gold dust for articles of clothing, knives, tobacco etc. For an old vest our hero bought a basket of oranges and for a handful of tobacco five large cocoanuts. This was a most valuable transaction for the lad inasmuch as, though he drew his daily allowance of tobacco, he had not acquired the habit of using it. The milk of the cocoanuts was highly appreciated when the *Siren* again ran short of water.

It was while the *Siren* was in St. Thomas, that Samuel had the first and only real occasion to complain of the tyranny of the petty officers in the American navy—and the style in which it was handled by Captain Nicholson is sufficient commentary of the humanity of our service in those days.

The petty officers messed by themselves and had a large, awkward boy, entirely unaccustomed to sea life, to wait on them. This led to some of the officers imposing upon him, even to the extent of knocking him around and using a rope's end on his back. For some reason Leech was ordered to take this boy's place and he, from the start, determined to resent this treatment.

One day the gunner came below for his share of whiskey and found it gone, his messmates having drank it all. He turned upon Samuel and asked for the whiskey. The lad boldly answered:

" I know nothing about it," upon which the gunner broke into a violent rage using the most improper language.

Leech at once went on deck and reported the matter to Captain Nicholson. The gunner was summoned and was warned that if he ever repeated the offense he would be punished. Leech had no further difficulty on that score.

Soon after this our hero had the satisfaction of playing a practical joke on this same gunner. Putting to sea the *Siren,* as usual, ran short of water (on account of the supply becoming foul) ; but the gunner, being " an old bird," had provided against such a contingency by having a keg of it securely locked in a room for his private use.

One hot night, when the throats of all were parched with thirst, Samuel met the boatswain's mate and said:

" If I were minded to play the rogue I could hook some water."

" Where? " eagerly asked the mate, who was almost dying with thirst.

" I have a key that will fit the lock of the room where the gunner keeps his water keg."

" Well," said he, " give me the key and I will be the rogue while you keep watch for the old sinner."

After drinking all they wanted of the delicious liquid, they locked the door and returned to their posts. The following day the gunner began throwing out hints, broadcast, about " sneak " thieving, what an unpardonable crime it was aboard a war ship and how he would just like to catch anyone doing such a thing aboard the *Siren*. Of course, no one had the slightest idea what he was raving about (excepting Samuel and the mate), the rest of the crew innocently supposing the want of water was driving the poor man out of his senses.

When Samuel and the mate next attempted to visit the keg, they found a new and stronger lock on the door.

CHAPTER XIV.

PRISONERS OF WAR.

Leaving St. Thomas, the *Siren* proceeded to Angola where she remained long enough to undergo a thorough overhauling and, after being cleaned and painted, she sailed for Boston— hoping to pick up a prize or two on the run across the Atlantic. So afraid was our hero of being retaken by the British that while at this place, he seriously considered the plan of deserting and finding refuge among the Africans. Better judgment prevailed, however, and he sailed with his ship.

Still, he employed every device to prevent recognition in case of capture. He allowed his hair to grow long but instead of tying it in a queue behind (a fashion then commonly affected by seamen) he trained it so it fell in ringlets about his face. This, together with several years growth, he hoped would prevent any of his former associates from recognizing him should he ever be paraded before them. He also adopted the peculiar dress affected by American seamen which was to open his shirt at the neck with the corners thrown back; on these corners being embroidered the stars of the American flag, with the British colors below.

Sailing from Angola, the *Siren* reached the island of Ascension in safety where she stopped long enough to examine the " post office." This was a box nailed to a tree near the shore where passing ships left letters and messages for other vessels to receive or carry to such different parts of the world as the directions called for.

Scarcely had the brig left this island when on July 12, 1814, the cry: " Sail ho!" arrested the attention of all on board. In a short time a large ship, which was taken for a merchantman, rose above the horizon but Captain Nicholson exercised great caution in his approach. He had no relish for placing himself in the dilemma of the bold Yankee privateersman who unhesitatingly ran under the guns of a 74-gun ship, believing that she was an Indiaman. He was not undeceived until he had called on her to surrender and the supposed Indiaman had run out a double row of huge guns.

" Oh! very well, then," smartly said the privateersman, " if you won't haul down your colors, I will."

It was to avoid such a mistake as this that Captain Nicholson approached the stranger with caution. It was soon discovered that she was under all sail making for the brig and, shortly afterward, it was seen that she was a 74-gun ship under English colors; upon which the *Siren* was promptly put about under all canvas to escape. Unfortunately the enemy had a wind most favorable for her and, as it was too evident that she was rapidly gaining, the Americans began to throw overboard their anchors, cables, hatches and, finally, their guns and ammunition in order to increase their speed. But the freshening breeze gave the huge seventy-four too much advantage and she was soon outcarrying the little brig and came lumbering down on her like an elephant after a spaniel.

Observing that his pursuer was almost within gunshot, Captain Nicholson ordered Quartermaster George Watson to throw the private signals overboard. "This," said Samuel, "was a hard task for the noble-hearted fellow. As he pitched them into the sea he said: 'Goodby, brother Yankee'; an expression which, in spite of the mortifying situation, forced a smile from the officers."

The report of a heavy gun now came booming through the air as a signal for the brig to heave-to or look out for the consequences. It was well that the *Siren* obeyed as promptly as she did for

MEDWAY CHASING THE SIREN.

they afterward learned that a division of the seventy-four's gun crews had strict orders to sink her if she made the least show of resistance. Heaving-to, Captain Nicholson caused the colors to be struck and waited while the enemy " came rolling down on us like a huge avalanche rushing down the mountain side to crush some poor peasant's dwelling."

Surrounded by his officers on the quarter deck, the British commander hailed:

" What brig is that? "

" The United States brig *Siren*," replied Captain Nicholson.

" This is His Britannic Majesty's ship *Medway*," he answered. " I claim you as my lawful prize."

Boats were now lowered and in a short time the Americans were transferred to the seventy-four; the officers being comfortably quartered with the British officers but the sailors were stowed away in the poky cable tier where they were formed in messes of twelve, each mess having an allowance for only eight men. This harsh treatment, in a short time, caused considerable suffering from hunger.

To Samuel, however, this was a small matter compared with his anxiety about the discovery of

his real character. When first going aboard the
seventy-four "the sight of the marines," he said,
"made me tremble for my fancy pointed out
several of them as having formerly belonged to
the *Macedonian.* I really feared I was destined
to speedily swing at the yard-arm."

On the day after the capture, all the prisoners
were marched to the quarter deck of the *Medway*
with their clothes bags to undergo a strict search ;
for the English knew that the *Siren* had just
come from the African coast and it was believed
that many of her crew had gold dust with them.
A most thorough examination was made, the men
being required to remove their outer garments
so as to facilitate the search. What little gold the
Americans had, was taken from them without
ceremony and appropriated by the officers of the
ship.

Arriving at Simon's Bay, the prisoners were
landed and were compelled to make the remainder
of their journey to Cape Town, twenty-one miles
distant, on foot. Leech recorded: "We were
received at the beach by a file of Irish soldiers.
Under their escort we proceeded seven miles,
through heaps of burning sand, seeing nothing
worthy of interest but a number of men engaged
in cutting up dead whales on the seashore.

"After resting a short time, we recommenced our march, guarded by a new detachment of soldiers. Unused to walking, as we were, we began to grow excessively fatigued and, after wading a stream of considerable depth, we were so overcome that it seemed impossible to proceed any further. We dropped down on the sand, discouraged and wretched. The guard brought us some bread and gave half a pint of wine to each man. This revived us somewhat.

"We were now placed under the guard of dragoons. They were very kind and urged us to attempt the remaining seven miles. To relieve us, they carried our clothes bags on their horses and, overtaking some Dutch farmers going to the Cape with broom-stuff and brush, the officer of the dragoons made them carry the most weary among us in their wagons. It is not common for men to desire the inside of a prison but we heartily wished ourselves there. At last, about nine o'clock that night, we reached Cape Town, having left one of our number at Wineburg through exhaustion, who rejoined us the next day. Stiff, sore and weary we threw ourselves on the hard boards of our prison where we slept soundly until late the next morning."

When Samuel awoke the following day he

found himself in a prison that had recently been occupied by several hundred American and French sailors. It consisted of a large yard, surrounded by high walls, strongly guarded by soldiers. Within this inclosure was a shed divided into three rooms; none of which had a floor, saving that afforded by Mother Earth. Around the sides of the shed were three rows of benches, one above the other and, by spreading their clothing on them, tolerably comfortable bunks for sleeping were formed. A few of the *Siren's* men, however, preferred to swing their hammocks; so accustomed had they become to that snug style of resting.

Most of the petty officers and soldiers were very kind to the Americans but, at first, several of them showed a disposition to be tyrannical. They were quickly cured of this by an ingenious Yankee device. Whenever one of these surly petty officers was on duty, the Americans bothered him by hiding so as to delay him in the morning and evening " round-up " of the prisoners. This protracted his time of duty when he was most anxious to be relieved.

Of course, the relief would not permit the former guard to go, until every prisoner had been accounted for. When several were missing, others were sent to find them and they, in turn, would hide and, so did the third batch of messengers. This vexatious delay sometimes kept the obnoxious tyrant an hour longer on duty. As these provoking delays occurred only when the objectionable officer was concerned, he soon came to understand it—and mended his ways accordingly.

Having triumphed over this annoyance, the Sirens next turned their attention to an old Dutchman named Badiem, who had the contract for supplying provisions for the prisoners. He had already found that it was dangerous to attempt cheating Yankees by supplying cheaper and poorer bread and so he was now more cautious.

It was not long before the Sirens found that a very inferior quality of bread was being furnished and, taking counsel among themselves, they decided to " fix " Badiem. According to British prison regulations, a superior officer was required to visit the prisoners every day and see to it that they were properly treated. This officer happened to be a kind old gentleman who had seen service in our war for independence and had been in the Battle of Bunker Hill—consequently he entertained the highest respect for Yankees.

"He had the profoundest respect for American character," said Leech, "and could not speak of the Battle of Bunker Hill without tears."

One day a friendly sergeant being on duty, the prisoners gave him a piece of the Dutchman's bread, complaining that it was not fit to eat. At the usual time the gallant old general, mounted on a fine, dashing charger, came round and asked the usual question:

"Everything all right?"

"No, sir," replied the sergeant.

"What is the matter?" asked the veteran.

"The prisoners complain of their bread, sir."

"Let me see it," commanded the general.

The sergeant gave him the piece. The general wrapped it carefully in a piece of paper, clapped spurs to his horse and galloped off. On the following day the prisoners had better bread than ever before and an order came for a man from each of the three rooms to go with the sentry every morning to Cape Town to examine their daily provisions—and if it proved not what it ought to be, to reject it at old Badiem's expense.

This upheaval of the Dutchman's dreams of ill-gotten profit, put him into the wildest rage but, so long as the Sirens were in that prison he never dared to again foist poor fare on them. Old

Badiem declared that he would rather feed one thousand Frenchmen than one hundred Yankees.

Leech records: " We now had an abundance of beef and mutton and a full allowance of bread. The mutton was excellent. Besides our prison allowance, we had an opportunity to purchase as many little luxuries as our slender finances would permit. These were furnished by a slave who was the property of the old Dutchman and who was so far a favorite as to be indulged in two wives and the privilege of selling small articles to the inmates of the prison."

For reasonable charges this sable polygamist provided coffee, tea, fish, sausages and fruit so that on Christmas Day the Sirens had some semblance of a jollification. In order to procure money for these luxuries, the prisoners were permitted to braid hats, make toy boats and such fancy articles as would sell in the town.

One day, Samuel became quite ill in the prison and his shipmates advised him to go to the hospital in Cape Town. It seems that he had been taken in a similar way when aboard the *Siren* and the surgeon had prescribed an ounce of salts that caused him the most horrible nausea. So, when the hospital was now suggested, Samuel at once associated it with those dreaded salts.

"I would go to the hospital," he said, "if I thought they would not give me salts."

His shipmates assured him that he would not receive such a remedy so, under the guard of a sentry, he sallied forth from the prison to the hospital.

"Well, my boy," cheerily asked the Doctor, "What is the matter with you?"

With many wry faces Samuel explained his symptoms, whereupon, the docter promptly turned to his assistant and said:

"Doctor Jack, six ounces of salts for this boy!"

Poor Samuel felt like jumping out of the window and he would have done so had he not known that a bullet would have overtaken him. An ounce of salts in the *Siren* had caused him excruciating agony—and now he was compelled to swallow *six!*

But there was no way of evading the dose. He took it and, much to his relief and surprise, found that they were an entirely different and much milder dose from that he had taken aboard the brig; the former being Epsom and the latter Glauber's salts.

So pleased was our hero with his trip to Cape Town and the opportunity to stroll about the

streets that, soon afterward, he feigned illness.
He repeated this once too often, however. The
Doctor seeing through the trick, gave him a dose
of medicine which cured Samuel of any further
desire to visit the hospital.

CHAPTER XV.

UNDER THE HALTER'S SHADOW.

There was a small prison at Cape Town called The Trunk, so dubbed on account of its scanty dimensions. To this place all the prisoners, where the *Siren's* crew was confined, were transferred when they became too refractory. Here they were kept on bread and water for such time as their " judge " deemed necessary. And it must be said that the Americans never complained when any of their number was thus punished— provided he was guilty.

One day, however, two of the *Siren's* crew were threatened with banishment to The Trunk most unjustly. It seems that two of the prisoners had washed their clothes and unwittingly had hung them over a line directly in front of the path leading to the prison doctor's office. Observing the clothing and being too proud to bend his head or go around the wet clothing, the doctor took out his knife and cut the line so the clothes fell in the dirt and were soiled.

The owners of the " insulted shirts and trousers " angrily inquired who had cut the line and were told that it was the English doctor. This brought forth a volley of sailor profanity that was not at all complimentary to the doctor who, overhearing it, ordered the two men to The Trunk.

The Sirens determined to resist and when the sergeant came to seize the men, all the Americans turned out in a body declaring that they would all go to The Trunk together. As the prisoners were in a state of mutiny, the guard was called out and ordered to load and fire. Upon this, the Americans shouted:

"Fire away! You will have but one fire and then it will be our turn!"

By this time all the broken bottles, stones and sticks in the yard had been picked up and the prisoners stood ready to open the battle.

Realizing that he would be overpowered, the sergeant recalled his men and the Sirens never heard any more about it—at least not from the enemy's side.

Such experiences afforded a welcome relief to the dreary monotony of prison life at Cape Town for the *Siren's* people had now been in "durance vile" seven months. Further excitement was caused soon after the "mutiny" by a midnight alarm.

One night, when all in the prison save the

guard were asleep, all hands were awakened by
the approach of a large party of shouting and
singing men and women, preceded by a band of
music. The prisoners turned out in a jiffy—
hoping that they were about to be liberated by a
daring raid from some American war ship. They
rushed, in a body, to the prison gate; ready to
perform their part in the rescue—if rescue it was.

Much to their chagrin—and afterward to their
no small amusement—they learned that the cause
of the commotion was a Dutch wedding party
going to the house of old Badiem, the prison
caterer, who lived nearby. The band of music,
very inappropriately, was playing that familiar
tune " A Free and Accepted Mason."

When the news of the capture and burning of
Washington reached this dreary prison, the Sirens decided to make a break for liberty. A
carefully laid plan to rise at night, overpower the
guards and proceeding to Simon's Bay, cut out
some ship and sail to America, was agreed to.
But through treachery it reached the ears of the
garrison, with the result that the guard was
doubled while the arrival of a company of dragoons rendered the project impossible of execution.

Shortly after this an English missionary, the
Rev. George Thom, asked permission of the

prisoners to preach to them on Sundays. Some
of the sailors objected on the ground that he
would laud the king but the prevailing sentiment
was " Let him come and show him that Americans know what good behavior is." They recalled the experiences of the eccentric Rowland
Hill, who, when attacked on a preaching tour,
was saved by a few sailors rallying about him
and dispersing the mob.

Cleaning up one of the rooms and arranging
benches, they welcomed Mr. Thom and his ami-
able wife on the following Sabbath: Instead of
preaching about kings and princes, as some of the
Americans feared, he gave them an earnest, simple
discourse which so pleased the men that they
invited him to come every Sunday. As some of
the hardy seamen expressed it: " He shot away
my colors," " He gave me a broadside " etc.

During the week Mr. Thom would visit the
prisoners, distributing healthful literature for
them to read. The result of this little thoughtful-
ness was most touching on these mariners. So
little accustomed were they to any consideration
or attention, they were readily led by the kindly
ministrations of this good man and his wife.
Gambling, profanity and other vices became un-
popular and were finally discarded altogether.

As some expression of their appreciation, the

Sirens presented Mr. Thom many gifts worked in a rough way by their hands—doubly precious to the good man on that account. One was the model of a full rigged ship, another was a hat made from bullocks' horns—the horn being pealed into narrow strips and woven together in shape for head gear.

About the middle of March, 1814, the 74-gun ship of the line *Cumberland* arrived at Cape Town to transfer the prisoners to England, preparatory to their voyage to the United States. While this was joyful news to most of the *Siren's* people, it was full of seriousness for our hero.

"The tidings filled me with fear," recorded Samuel. "Directly to America I would gladly have gone but to be carried to England, in one of her ships of war, was like going to certain death. How was it possible for me to escape detection? How could I avoid meeting some of the old Macedonians who would, of course, recognize and betray me?

"These questions tortured me beyond endurance and almost induced me to volunteer to remain at the Cape. I felt like an escaped criminal, with the officers of justice at his heels. Death at the yard-arm haunted me day and night. No one can imagine my uneasiness unless he has been similarly placed."

After the usual delays, the Sirens were stowed aboard the *Cumberland* in far more comfortable quarters than they had had in the *Medway* for, instead of the stuffy cable tier, they had roomy bunks on the upper gun-deck; besides which they had plenty of good food.

Arriving at St. Helena, part of the prisoners were transferred to their captor, the *Medway*, and the rest to the 50-gun frigate *Grampus;* young Leech being sent to the latter. "This transfer to the *Grampus* greatly alarmed me," said Samuel, "since the more men I saw the greater was the chance of detection. I had already escaped being known on board of two seventy-fours with their half thousand men each but I could not promise myself the same immunity much longer. However, as I saw no face that was familiar when I went on board, I felt a little more at ease."

That night, however, proved to be one long remembered by our hero as one in which he grew several years older in as many minutes. About nine o'clock, when every thing in the great frigate had settled down to the orderly quiet of the hour, a call from the officer of the deck was passed along the main deck which, in the stillness of the night, sounded as if it had come from a speaking trumpet. It was:

12

"Pass the word for the boy Leech!"

For a moment Samuel's heart stopped beating and then began thumping like a trip hammer as, in agonizing fear, he awaited the outcome of this summons for "the boy Leech"—which was always the way he had been summoned when aboard the *Macedonian.*

"Pass the word for the boy Leech!" was repeated by several gruff boatswain's mates and, as each order came nearer to the place where our hero was waiting, there seemed to be no doubt but that it was intended for him. Indeed, several of the American prisoners said to him:

"That means you."

Samuel was so terrified that he could make no motion, nor could he control his voice so as to answer. Perhaps it was fortunate for him that he remained silent for, a moment later, he heard some one say:

"Your master wants you."

This convinced Samuel that there was a "Boy Leech" in the frigate's crew as well as among the prisoners and so it proved—though our friend declared that he did not breathe freely again for a week afterward and that at night he suffered from the most hideous nightmares.

When at St. Helena it was learned that war with France was ended. This was gratifying

VIEW OF A GUNDECK IN AN ENGLISH LINE OF BATTLE SHIP.

From a photograph.

news to the sailors as they were all hoping to get discharged. On the run from this island to England, however, it was learned from a passing ship that Napoleon had escaped from Elba and was at Paris with sixty thousand men so that the war was on again. "Nothing," said Samuel, "could exceed the joy of the officers at this announcement and the corresponding vexation of the crew. The former dreaded peace because it meant half pay, no prize money, and little chance of promotion."

At last the white cliffs of Old England rose above the horizon. To avoid suspicion our hero pretended to be very much interested in every thing about the "new" land, asking such questions as any foreigner might. "I could not behold myself approaching my native land," records Samuel, "without many misgivings. To a man who knows a halter is hanging over his head, everything furnished cause for alarm; a piercing look, a whisper or the sudden mention of my name caused me to tremble."

One day, before they made port, Captain Nicholson inadvertently came near disclosing Leech's secret before the officers of the *Grampus*. It seems some discussion had been going on between the American and British officers when Nicholson sent for Samuel to clear up a point involving Salem; from which place, the American officer all along had supposed the boy really came.

Appearing before the officers in fear and trembling he was asked a question about Mr. Crowninshield of Salem. Fortunately, Leech knew of him and answered satisfactorily. It was with immeasurable relief that he found no other questions were to be asked for every moment Samuel feared that he would get into the same trouble he had relative to his "native city of Philamadelph."

Nor was this an idle fear that Leech had, about being betrayed unintentionally by his friends. A case occurred only a short time before, in the very port for which the *Grampus* was making, where a mother innocently revealed the identity of her son. Not knowing that he was a deserter from the royal navy, she went aboard a newly arrived English man-of-war and asked for him, giving his name and rating. They replied that no one of that name was aboard.

"He is among the Yankees," unthinkingly remarked the good mother.

There happened to be some American prisoners and, hearing the remark, an officer summoned up the prisoners and paraded them on the gun deck. Seeing her son among the number the poor woman exclaimed:

"Oh, Tom! I have brought you a clean shirt!"

The officer, who was standing by, then stepped up to the man and said:

"It's a clean shirt you want, is it? I'll give you a clean handkerchief"—meaning that he would be hung. The unfortunate lad was at once placed in irons, in the presence of his mother. A courtmartial was held and on the following day he was hung at the yard-arm. It was fortunate for our hero that Captain Nicholson did not question him too closely about Salem in the presence of the British officers.

Arriving at Spithead the Americans were transferred to the prison-ship *Puissant*, a war vessel which had been captured from the French. "Here we were treated with great leniency," records Samuel. "We were even allowed liberty to go ashore. Had I dared, I would have run away—but the dread of a halter restrained me. I did not even venture to write to my mother, lest she should be tempted to visit me, or even write, as a letter from any place in England might awaken suspicions as to my true character and she might share the grief of the too-fond mother who innocently sent her son to the gallows."

After a stay of several weeks in the *Puissant*,

the Sirens were transferred to the gun-brig *Rover*
which was to transfer the prisoners to Plymouth,
England.

It seemed a part of the Admiralty's policy to
transfer American prisoners from one British
war ship to another as many times as possible—
evidently with the object of increasing the chances
of detecting deserters. So far, our hero had
passed under the surveillance of the English in
the *Medway*, in the prison at Cape Town and
of the people in the *Cumberland, Grampus, Puissant* and *Rover*—numbering in all, several thousand British officers and sailors; certainly an
ordeal which nothing but Providence enabled
him to pass through without detection.

But a still severer ordeal was awaiting him.
In his diary Leech records in reference to his
induction on the *Rover:* "Here was a double
risk again before me. I had a risk of being
known by the crew of the *Rover* and by the many
people who had known me at Plymouth during
my previous stay at that port. However, the
good hand of Providence was with me to pre-
serve me. We reached our port in safety where,
to our great delight, we heard that the *Woodrop
Simms,* Captain Jones, of Philadelphia, was to be
the cartel to convey us to America."

Before the Americans were permitted to tread

her decks, however, the prisoners (again following the policy of the Admiralty for detecting
deserters) were sent aboard the *Royal Sovereign*
where they would be exposed to the gaze of eight
hundred men—a large bounty being offered to
the man who would reveal the identity of British
deserters. Unfortunately for Samuel, this *Royal
Sovereign*, on a previous occasion, had sailed
in company with the *Macedonian* and Leech was
known to many of her people.

To avoid recognition our hero resorted to a
stratagem. He says: "Whenever any of her
men came near our quarters, I endeavored to
look cross-eyed or closed one eye so as to appear
partially blind; and in various other ways altered
my appearance so that even an old shipmate
would have been puzzled to recognize me at first."

CHAPTER XVI.

HOMEWARD BOUND.

At last the grateful news that the *Woodrop Simms* was ready, reached the prisoners and in August, 1815, Samuel went aboard her the happiest boy that ever breathed the breath of life. Here he met a number of other Sirens who— during their short stay at Plymouth—had been confined in Dartmoor Prison.

It would be supposed that these men, after such a protracted incarceration in British prisons, would be without money. At least, so argued Samuel. What must have been his surprise, then, when he saw these same Dartmoor prisoners purchasing large quantities of luxuries for their voyage across the Atlantic—and they had the coin to pay for them, too.

Later on, Samuel learned that this was counterfeit money which these prisoners had made during their involuntary stay at Dartmoor. How far this spurious money circulated before the authorities learned of it, Samuel does not state, neither did he care for on the following day, the *Woodrop Simms* set sail for America.

It was one morning when the cartel was well on her way across the Atlantic that the sequel to Captain Nicholson's near approach to a betrayal of Leech's identity before British officers in the *Grampus,* came about. Nicholson again asked Samuel something about Salem. Our hero gave a knowing laugh. The American commander, somewhat indignant, asked:

"Why this levity?"

"Sir," replied the youth, "Salem is not my native place by a considerable."

"What do you mean?" inquired Nicholson, somewhat mystified by Samuel's manner.

It was then that Leech, for the first time, told Nicholson how he had been captured in the *Macedonian,* had deserted and had shipped in an American war ship and had, all these months been under the shadow of a halter. Captain Nicholson warmly congratulated the lad on his many narrow escapes.

As showing the fickleness of the old-time sailor's whims, the following extract from Samuel's diary will be given: "During the voyage a great deal was said about quitting the sea and settling down in quietness on shore. One of our shipmates, named William Carpenter who belonged to Rhode Island, had a particular enthusiasm for farming. He promised to take me with him

where I could learn the art of cultivating the
soil. Many of us made strong resolutions to
embark in some such enterprise. The pleasures
of agriculture were sung and praised among us
in so ardent a manner that he must have been
incredulous indeed who could have doubted for a
moment the certainty of quite a number of our
hands becoming farmers whenever we should
gain land."

One night as they swung in their hammocks,
talking with great earnestness about their favorite theme (farming), the wind blowing quite
freshly on deck, one said:

"If ever I get home, you won't catch me on
board of a ship again."

"Yes," said another, "farmers live well at
any rate. They are not put on an allowance but
have enough to eat. If they work hard at it all
day, they can turn in at night—and if it blows
hard the house won't rock and there are no sails
to reef."

While these and other good resolutions were
being formed, the wind began blowing harder
and harder. From occasional puffs it quickly
grew to a tremendous gale. Realizing that they
were in for a storm and thinking that all hands
might be required, those below went on deck to
assist. It was now blowing a hurricane, the

wind howling and whistling through the rigging,
the wilder roar of the angry sea, the shouting of
the officers and the intense darkness all conspired
to present a scene of indescribable terror.

Just as our hero stepped on deck, a heavy wave
broke over the cartel, drenching all hands and
threatening to carry her down. Shortly afterward the crash of a falling mast was heard. It
was a topmast going overboard, leaving a yard
in the slings. There were so many men on deck
now that they only encumbered each other's
movements; so some went below with the full
expectation that the ship would founder before
morning; and, with true sailor-like philosophy,
they argued that it would be as well to drown in
their hammocks as on deck.

During this night of appalling danger the men
manifested curiously varying symptoms of alarm.
"Some prayed aloud," records Leech. "Others
cursed as if in bravado shouting 'We are all
going to perdition together!' For my own part
I kept repeating the Lord's prayer and renewing
those promises, so often made in moments of
apparent destruction."

At length day broke, revealing the sad havoc
made by the storm. The shattered state of the
cartel's masts and rigging reminded Samuel of

the condition of the *Macedonian* after her action
with the *United States,* excepting that there were
no wounded or dead encumbering the deck.
Captain Jones, who commanded the *Woodrop
Simms,* declared that, though he had been at sea
twenty-five years, he had never experienced such
a frightful storm. He had not left the deck all
that night. Fortunately for all hands, the ship
was nearly new, exceptionally strong and an excellent sea boat.

As the gale abated, repairs were made and the
cartel proceeded on her voyage, meeting a number of vessels that had suffered even worse than
she, while (as they afterward learned) many
craft went down with all hands on board. That
gale was on the 9th and 10th of August, 1815,
and was long remembered among sailor folk as
the worst in their experience.

When so many resolutions about " turning into
farmers " had, apparently, been " clinched " by
this close swish against the winding sheet of
Death, we would expect that the men making
them would exert all effort to get into the rural
districts as soon as possible after reaching port.
Such, however, did not prove to be the case.

Arriving in New York, safe and sound, they
were paid off and, in a twinkling, the hardships

AN AMERICAN 44-GUN FRIGATE RIDING OUT A GALE.
From the original painting.

and perils of the sea were forgotten as most of
the sailors plunged headlong into dissipation
which was continued so long as their money held
out. As Samuel well expressed it: "We felt as
if New York belonged to us and that we were
really the happiest, jolliest fellows in the world."

It is only in justice to our hero to say, however,
that at first he made honest effort to find employment on shore. He looked up the bootmaking
establishment in Broadway, where he had begun
an apprenticeship, and was truly disappointed on
learning that his employer had moved to Philadelphia. After this damper on his good resolutions our hero, we fear, did not make other
serious attempts in this line. He had a hundred
dollars in his pocket and was content to float
about the city until that was gone when, like most
of his companions in the cartel, he enlisted again.

Samuel shipped in the United States war brig
Boxer which had been captured by the *Enterprise*, September, 1813. The *Boxer* was now
commanded by the celebrated David Porter who
was captain of the *Essex* in her famous cruise
in the south Pacific, 1812-1814. "Although Captain Porter was stern and severe," said Samuel,
"he never used bad language. He always spoke
with the utmost deliberation but with such ob-

vious feeling that we often trembled to hear his voice."

While in the *Boxer*, Samuel learned a new "trick" in maintaining discipline which was far more effective and less brutal than flogging. As it was now in time of peace, night watches were prone to steal a doze while on duty.

To check this habit, Captain Porter ordered that any man caught sleeping was to be aroused by a handspike—not too gently applied. The offender was then obliged to take the handspike and hold it in his hand as a badge of disgrace throughout the watch, unless he discovered another man asleep when he was to awake him in a similar manner and pass the handspike to him. By this simple means, the night watches were sufficiently "interested" to keep awake.

After his service in the *Boxer*, Samuel, now fourteen years old, renewed his resolutions about seeking employment on shore and now, without funds, he set out in the dead of winter and painfully made his way—ragged, footsore and cold— to New Haven, Hartford, Coventry and Mansfield. At the last town he met an old shipmate in the *Macedonian* who had wandered into Connecticut, married and was in fairly comfortable circumstances. Through his influence Samuel found steady work and attending a Methodist revival, he became a staunch supporter of that faith the remainder of his days.

Our hero describes how he was induced to attend the revival by some " worldly young men " who thought to have sport with him: " Some of the young men who spent their evenings with me listening to my sea yarns invited me to attend a meeting of the Methodist Church. But they greatly misjudged the character of seamen. I attended the meeting but not to make sport. One Sabbath evening my friend Ella Dunham asked:

" ' When do you intend to set out and seek religion?' I replied, somewhat evasively:

" ' Any time.'

" ' Well' said he, ' are you willing we should pray for you and will you go forward for prayers to-night?'

" To this I replied that I would think of it. The meeting proved to be intensely interesting. My desire to express the inward working of my mind grew strong. . . . . I determined to rise and speak though the Evil One whispered ' Not yet, not yet' in my ears.

" Just as I stood up, some one (not seeing me) began to sing but my friend Dunham checked the singing ' because a young man wished to speak.' He had seen my move. Thus encouraged I told them I was then nineteen years old and it seemed to me too much of life to spend in sin, that eternity was a solemn idea and I desired them to tell me how to enter upon it with joy. They proposed to pray with me. We then all kneeled down together. Most fervently did they pray for the divine blessing to rest on the stranger youth, bowed in penitence before them and most sincerely did I join my prayers with theirs before the throne of God."

Samuel records that of all his shipmates who survived the naval battle of October 25, 1812, he knew of only one, besides himself, who embraced religion; and his name—singularly enough—was John Whiskey.

In all his wanderings Samuel had not forgotten his mother but, owing to his frequent changes of address, he did not get a letter from her until he had settled in Connecticut—some eight years after leaving Blenheim. How eagerly the good woman had followed the fortunes of her son is revealed in the opening words of this, her heart's message: "My dearly beloved Child. I cannot describe the sensations I felt when I received a letter from your dear hands. It was the greatest pleasure I have enjoyed since you left me. . . . . I sent your letter to Lady Churchill, formerly Lady

Francis Spencer. Both Lord and Lady Churchill were glad to hear from you and are your well-wishers. The Duke and Duchess of Marlborough are both dead. . . . . Lord Francis makes a very good master."

Much as Samuel desired to visit England and again see his beloved mother, he was deterred from so doing on account of his desertion from the British navy. This fear was enhanced by a conversation he had with the traveler, Lorenzo Dow, who assured Samuel that he had recently seen four men hanged in England for just such an offense as our hero had committed.

Samuel then endeavored to persuade his mother and his step-father to emigrate to America but, before his letter reached England, Mr. Newman died and Mrs. Newman felt that it would be unwise to migrate at her time of life.

Soon afterward Samuel left his employer in Mansfield and, purchasing a horse and wagon, started into business for himself; traveling from town to town, selling steelyards etc. Accumulating a small capital in this way he opened a store in Mansfield.

Like nearly all New England stores in those days the one opened by our hero had a corner where intoxicating liquor could be obtained. But

13

soon afterward, hearing a temperance sermon by
Dr. Hawes of Hartford, he gave up that lucrative
branch of the business. He said: "I could hold
out no longer, in spite of the example of our best
citizens (some of whom often drank, though spar-
ingly, at my house) I gave it up. . . . . I have ever
regarded that act as among the best of my life."

In time, Samuel married a member of the
Methodist church and went to Somers, Conn.,
from which place he soon moved and finally
settled in Wilbraham, Mass. Here the years
glided pleasantly and swiftly as our hero pros-
pered in business and grew in the esteem of his
fellow townsmen.

A matter of business calling him to New York,
Samuel learned that the *Macedonian* was in that
port; and, with true sailor-like attachment for
the "old ship," he visited her. He records: "I
stood on the spot where I had fought in the din
of battle and, with many a serious reflection, recalled the horrors of that dreadful scene. The
sailors, on witnessing the care with which I
examined everything, and supposing me to be a
landman, eyed me rather closely. Seeing their
curiosity, I said:

"'Shipmates, I have seen this vessel before
to-day; probably before any of you did.'

" The old tars gathered around me, eagerly listening to my tale of the battle and they bore patiently and with becoming gravity the exhortation to lead a religious life with which I closed my address."

CHAPTER XVII.

AGAIN AT BLENHEIM.

As has already appeared in these pages, Mr. Leech—we must now be more careful in mentioning him for, not only did he have a wife and three children but he had attained the proud distinction of being a sovereign American citizen —had long intended to make a visit to England to see his mother and the scenes of his boyhood again, but had been deterred from so doing by fear of death for desertion from the British navy.

To remove this obstacle his mother, through the influence of Lady Churchill, secured the following official protection for our hero:

LOWER BROOK STREET, Nov. 7, 1821.

MRS. NEWMAN:

I consulted my brother William upon the subject on which you wish for advice, as neither Lord C. nor myself could undertake to answer your inquiry; and I am glad to hear from him the following explanation in reply: "There is nothing to prevent Mrs. Newman's son from coming home; for when the war was terminated, he was safe, even if he had entered the enemy's service; but he will, of course, forfeit the pay and any prize money due him."

I am, much yours,

F. CHURCHILL.

While he was unmarried, a trip to England and back would have been an easy matter for our hero but now, that he was a man of family, the items of passage money and incidental expenses, assumed formidable proportions. It was this serious aspect of the undertaking that induced Mr. Leech to urge his mother to emigrate to America, where he promised her a warm welcome and a comfortable home. But the good woman was too deeply attached to British soil to be up-rooted in her old age and, between urging on one side and entreating on the other, the years flew by.

At last Mr. Leech received a letter from his mother which decided him to undertake the voyage; so, one pleasant morning early in June, 1841, the Leech family, formidably protected with baggage, set out from Wilbraham and, after a pleasant passage by way of Springfield and Hart-ford, arrived in New York where they took passage in the "splendid packet-ship *George Washington,*" bound for Liverpool.

After a "quick run of twenty days," they arrived at Liverpool and it was here that our hero's well engrafted Americanism began to assert itself. Having bestowed due praise on the magnificent docks, he had to contrast "the dark, dingy aspect of Liverpool, everywhere discolored by the fumes of coal-smoke, with the light, cheerful aspect of our American cities; and giving preference to the latter, notwithstanding my English prejudices."

After passing through the charming rural scenery between Liverpool and Stafford, and dilating on the risk he ran of "being torn asunder by the eagerness of two hackmen who, as we were the only passengers left at the station, were especially zealous for our patronage," Mr. Leech and family were soon deposited at the door of his sister, the wife of William Tills.

"Although I had not seen her for thirty years yet, no sooner did she see me than, throwing her arms around my neck, she exclaimed: 'Oh, my Brother!' I need not add that our reception was cordial and our stay with them characterized by every trait of genuine hospitality."

Here, also, our hero showed his good citizenship when he records: "Having been so long away from England, everything peculiarly English struck me with almost as much force as it would a native American. Hence my feelings revolted at the sight of innumerable beggars and vagrants, who crowded the streets; and houseless families imploring a crust for their half-naked little ones with many a tale of sorrow, were sights that greeted my eyes and pained my heart every day."

Remaining a week with the hospitable Tills our American friends proceeded to Birmingham and thence by stage to Woodstock, " subject, however, to the incessant exactions of a host of waiters, guards and coachmen who all wanted to be ' remembered.' "

" Oh, I'll never forget you," was Mr. Leech's vexed reply.

" The country through which we rode was delightful," continued our hero.

About dusk, the stage drew up with the usual bustle of importance and ceremony at the Marlborough Arms, the same hostlery from which our hero had departed, thirty-one years before, in search of fame and fortune. " The first object that met my eye," he wrote, " was the revered form of my mother, waiting on the sidewalk, eager to embrace her much-loved but long absent son. Springing to the ground I felt myself locked in her fond embrace. That was a moment of exquisite enjoyment. . Although my mother was deeply moved, she maintained a calm dignity

of manner. In a few moments she was showing
the way with the agility of a young woman.
She held a new-found grandchild at each hand.
Reaching her residence, which was near by, we
were soon seated around the well-loaded board,
the happiest family party in the world."

While at Blenheim, Mr. Leech paid a visit
to Oxford and saw the room in which John
Wesley studied and other places associated with
the birth of Methodism. Returning to Bladen,
our hero made preparations for the return trip to
America. He records: "After receiving the
utmost kindness, hospitality and evidences of
friendship from my family and friends, I took
leave of them forever.. Many of the neighbors,
with my mother, accompanied me to Woodstock.
There I wished her adieu and, while the coach
whirled away, she stood following us with her
eyes, the last of the company, until a projection
of the park wall hid us from each other—and
forever. Who could forbear a tear in such a
moment? I could not, and suffered the big drops
to roll down my cheeks at will."

Passing on to London, Mr. Leech visited Wanstead and learned that his aunt—she of the
twenty-two sons and two daughters—was yet
alive. Our hero says: "My brother asked her
if she knew me. Peering through her spectacles

and summoning up the imagery of the past, she at length called to mind her former *protégé* and clasped me to her arms with evident gratification."

On the 25th of August Mr. Leech and family sailed from London " in a fine ship commanded by Captain Eldridge, bound for Boston," and after a most tempestuous passage of *seventy-five days,* arrived at their destination. Returning to Wilbraham, " we were hailed," says Mr. Leech, " with joyful congratulations by our neighbors who had begun to mourn us as among the lost at sea."

Shortly after his return to America our hero received a letter announcing the death of his mother. Samuel Leech died in the year 1848 in Wilbraham.

INDEX.

THE END.

BY EDGAR STANTON MACLAY.

A HISTORY OF THE UNITED STATES NAVY,

From the earliest times to 1902 in three 8vo vols. with a total of 1832 pages and 170 maps, diagrams and illustrations. Index. Third edition published in 1902. $3 a volume.

This monumental work, from the time it was first published, 1894, has been pronounced by the leading literary critics of America, Europe and Japan as THE STANDARD.

Extract from chapter on "Battle in Mobile Bay," vol. II, pp. 425-426: "The view of the battle obtained from the tops of the National vessels was one of appalling grandeur. To windward the fleet and harbor were spread out in a beautiful panorama, the crews being distinctly seen firing and reloading their guns, while officers stood at the back of their men to see that there was no flinching, and others ran to and fro shouting orders in their endeavors to prevent a collision. To leeward dense volumes of smoke, illuminated by the rapid flashes of guns, partly obstructed the vision, but in the occasional rifts a tall mast with men in the rigging and with Old Glory still flying in the breeze would be revealed. Above all rose the dreadful roar of the tremendous cannonading, whose sharp impact upon the ear, giving the peculiar sound of shotted guns, seemed to come from all quarters with deafening rapidity, while the ships and their masts quivered like aspens from the recoil of their murderous broadsides.

"A glance below on the deck of the *Hartford* revealed the men in their different capacities, some loading and aiming the guns, some bringing up ammunition and others carrying down the wounded; but all stimulated to their utmost exertions by the ever-vigilant officers. Most of the men were stripped to the waist, many of them smeared with the blood of shipmates whom they had carried below. Others, although wounded, refused to go below and remained on deck fighting.

"What a pandemonium! What a hell upon earth! Shot, shell, grape, shrapnel and canister. How they shriek! How the men fight! dragging dead or wounded shipmates away so as not to encumber the guns. Bloody and blackened with burned powder, the perspiration running down their bodies revealing streaks of white skin, causes them to look like fiends. The sight of their fallen shipmates arouses the brutish thirst for vengeance and they load and fire with muttered imprecations on the enemy. Their officers walk among them with ' Steady, boys!', 'Take your time!', 'Be sure of your aim!', 'Let each shot tell!'

"In the midst of all this uproar stand Drayton and his executive officer, Kimberly, the latter smiling and twirling his goatee; both as cool as if ''twa a daily drill.' It was in reference to the heroism of the crew that Brownell wrote:

"'But ah, the pluck of the crew!
"'Had you stood on that deck of ours
"'You had seen what men may do.'"

BY EDGAR STANTON MACLAY.

A HISTORY OF AMERICAN PRIVATEERS,

Uniform with and forming vol. IV of Maclay's History of the U. S. Navy, 8vo, 559 pages, 37 specially prepared maps, diagrams and illustrations, index, published 1898—$3.

It stands alone as a remarkable story of American daring, enterprise and consummate pluck and nautical skill.

Extract from chapter on "An Escape From Mill Prison" (England), pp. 160-161: "On leaping over the hedge he [Lieut. Joshua Barney, U. S. N.] found himself in the superb private grounds of Lord Edgecombe. Wandering about in search of the servants' house, he was discovered by the gardener, who was much incensed by the intrusion. Barney pacified him by explaining that he had injured his leg and was seeking the shortest way to Plymouth. Giving the gardener a tip, Barney was conducted to a private gate opening on the river and hailing a butcher who was going by in a small wherry with two sheep to market, our adventurer got aboard. By this means Barney avoided the necessity of crossing the river by the public ferry, and also that of passing by Mill Prison and a chance of meeting the guard.

"Immediately on receiving the report of the privateer's commander, Admiral Digby caused an inquiry to be made in all the prisons and places of confinement in or near Plymouth, and at the time Barney was sliding down the rope over the privateer's stern to get into a boat, his escape from Mill Prison was discovered; and at the moment he passed through Lord Edgecombe's private gate to the riverside, the tramp of the soldiers—all of whom were familiar with Barney—was heard, passing the very hedge he had just vaulted over, on their way to take him back to prison.

"That night Barney gained the house of the venerable clergyman that he had left only the morning before. The same evening Colonel Richardson and Dr. Hindman arrived at this house also, having been released from the privateer after the guard from Mill Prison had inspected them. While these fugitives were seated at supper, laughing and joking over their hapless adventures, the bell of the town-crier was heard under the windows and the reward of five guineas for the apprehension of 'Joshua Barney, a rebel deserter from Mill Prison,' was proclaimed. For a moment. . . . . ," etc.

REMINISCENCES OF THE OLD NAVY,

From the journals and private papers of Captain Edward Trenchard, U. S. N., and Rear-Admiral Stephen Decatur Trenchard, U. S. N., 8vo, 372 pages, index, $2.50.

It gives fascinating " inside " glimpses of our navy from 1800 to 1883.

Extract from chapter "On The West Coast of Africa," pp. 18-20: "The musicians of the high seas in those days [1820] did not occupy

14

the important position they hold in firstclass cruisers to-day, and the few lone and lorn manipulators of wood and brass in the *Cyane*, in all probability, would have made a poor showing in the highly cultivated musical ear of the modern Jack Tar. In fact, musicians at that time were regarded with condescending contempt by the hardy sailors as being, perhaps, good enough to tickle the ear with their tingling notes or to twitch a few muscles of the limbs into a jigging mood in fair weather; but when it came to real work, and an enemy was to be fought, they were fit only to be stowed away in a cable tier. But however that may be, the Jack Tar of that day had not the high musical mind of his descendants of to-day and he undoubtedly tolerated the tingling brass and the wheezy wood with his usual, good-natured indifference.

"The natives on the coast, however, were immensely impressed with the *Cyane's* band. To hear it was one of the events on the West African coast; and its fame extended even to the islands of the seas as the following incident will show: On May 19, 1820, the *Cyane* put into Port Praya, after a cruise in search of slavers. The fame of her band had preceded her for scarcely had she dropped anchor when a messenger came aboard with the announcement that 'His Excellency, the Governor-General, solicits the pleasure of Captain Trenchard's company, with that of all the officers of the *Cyane*, to tea this evening and would be highly gratified with having a few tunes from Captain Trenchard's band, which he solicits may be permitted to come on shore with their musical instruments, as the evening will be rendered delightful and pleasant by a full moon.' This enchanting invitation to 'tea' and a 'full moon' was sent through one Hodges, an English-speaking person on the island—and the above is as near a literal interpretation as can be given.

"Captain Trenchard complied with the request and attended the Governor. After a decent amount of time had been allowed the American officers for the contemplation of the tea and the full moon, the natives were treated to, what to them were, the awe-inspiring sounds from the *Cyane's* band. With forethought bred by experience, the band-master labelled in advance the tunes that he was about to render so that the audience would have no difficulty in knowing what melody they were 'feasting' upon. So impressed were the natives by this revelation of sound that the Governor, on the following Sunday, when the moon had again recovered her position in the heaven after her fullness, invited Captain Trenchard and Lieutenant Mervine 'and any of the officers that can be spared from the ship' to dinner—and, of course the band had to be exhibited again."

NAVY BLUE COMPANY
GREENLAWN, N. Y.

BY EDGAR STANTON MACLAY.

ADVENTURES OF REAR ADMIRAL PHILIP, U. S. N.,

From the diary and private records of the late Rear-Admiral who said: "Don't cheer, men, those poor devils are dying!" 12mo, 288 pages, $1.50.

Extract from chapter "At Annapolis," p. 58: "One of the instructors at Annapolis during Jack's novitiate was a greatly beloved man whose only fault—so the middies declared—was that of stammering when unusually exicited or nervous. One beautiful spring morning, when the middies were drilling in infantry tactics under the care of this officer, the youngsters were marching toward the seawall and were within a few feet of it, when their commander endeavored to give the order 'Halt!'

"The middies heard the hissing and spluttering noises behind them and knew perfectly well what the instructor was trying to say but, in that spirit of mischief so natural with boys, they marched right over the seawall and had waded into a considerable depth of water before the instructor finally gave vent to the word 'Halt!'"

MIDSHIPMAN PHILIP.

MOSES BROWN, CAPTAIN, U. S. N.,

Captain Brown rendered conspicuous, though forgotten, service on the ocean in the Revolution and the war against France. 12 mo, 10 illustrations, index, 220 pages, $1.50.

Extract from the chapter "A Prisoner of War," pp. 93-95: "Three days after her capture of the *George,* the career of the privateer [*General Arnold*] was cut short, she being captured [June 1, 1779] by the 50-gun ship *Experiment,* Captain Sir James Wallace. It is related that when Captain Brown gained the deck of the *Experiment,* Sir James asked him if he was the 'Captain of that rebel ship.' Brown replied:

'I was very lately; you are now,' and offered to surrender his sword. Captain Wallace refused to receive it, saying: 'I never take a sword from a brave man.'

" Sir James continued to extend every courtesy to his prisoner, treating him more as a guest. Taking Captain Brown into his private cabin, where he met other officers of the ship, a general conversation followed (over the traditional 'glass of wine') upon the affairs of the two countries, when Sir James proposed as a toast 'His Majesty, K i n g George the Third.' It was rather hard for the doughty Yankee skipper to accept but he swallowed his wine without remark. Sir James now called on Brown for a return toast—thinking, from Brown's silence that he had acquiesced in the sentiment and would respond with something of the like.

MOSES BROWN.

"Rising with much dignity and unawed by his position as a prisoner aboard a powerful enemy's warship, Captain Brown gave as a toast: 'His Excellency, General George Washington, the Commander-in-chief of the American forces.'

"The glass which Sir James had raised to his lips was hastily lowered and, turning fiercely on his prisoner, he asked: 'Do you mean to insult me, sir, in my own ship by proposing the name of that arch rebel?' 'No,' replied Captain Brown. 'If there was any insult it was your giving. as a toast, George the Third, which, however, I did not hesitate to drink to, although you must have known it could not be agreeable to me who, at this moment, am a guest although a prisoner.' Sir James at once perceived that if there had been a breach of etiquette, he had led the way and, like the honorable man he was, he suppressed his anger and drank to that 'arch rebel' Washington."

NAVY BLUE COMPANY
Greenlawn, N. Y.

BY EDGAR STANTON MACLAY.

JOURNAL OF WILLIAM MACLAY, Edited by Edgar Stanton Maclay,
William Maclay, with Robert Morris, represented Pennsylvania in
the first United States Senate, 1789-1791. 8vo, 452 pages, index,
$2.25.

> Maclay's successful fight against the introduction of monarchial
> forms in the "new" Government won for him the title of
> "Father of the Democratic party." A "canopied throne" and
> the title of "His Elective Majesty" for the President would
> have been given had it not been for the sturdy opposition made
> by Senator Maclay. His journal is the only connected record of
> the first United States Senate.

Extract from pp. 73-74: "Dined this day [June 11, 1789] with Mr.
Morris [Senator from Pennsylvania]. Mr. Fitzsimmons and Mr.
Clymer [Representatives from Pennsylvania] all the company, except
Mrs. Morris. Mrs. Morris talked a great deal after dinner. She did
it gracefully enough, this being
a gayer place and she being here
considered as at least the second
female character at court. As
to taste, etiquette etc., she is
certainly first. I thought she
discovered a predeliction for
New York but perhaps she was
only doing it justice, while my
extreme aversion, like a jealous
sentinel, is for giving no quarter.
I, however, happened to mention
that they were ill supplied with
the article of cream. Mrs.
Morris had much to say on this
subject; declared they had done
all they could and even sent to
the country all about, but that
they could not be supplied. She
told many anecdotes on this subject; particularly how, two days
ago, she dined at the President's. A large, fine-looking
trifle was brought to table and
appeared exceedingly well indeed. She was helped by the
President but on taking some

WILLIAM MACLAY.

of it, she had to pass her handkerchief to her mouth and rid herself of
the morsel; on which she whispered the President. The cream of which
it is made had been unusually stale and rancid; on which the General
changed his plate immediately. 'But,' she added with a titter, 'Mrs.
Washington ate a whole heap of it.'"

BY EDGAR STANTON MACLAY.

PENNSYLVANIA STATE
MONUMENT TO
SAMUEL MACLAY.

DIARY OF SAMUEL MACLAY, Edited by Edgar Stanton Maclay,

Being an account of the first official survey of the West Branch of the Susquehanna, the Sinnemahoning and Allegheny rivers in 1790. 8vo, 63 pages, $2.

Samuel Maclay, a brother of U. S. Senator William Maclay, was United States Senator from Pennsyvania from 1803 to 1809. He was a Congressman from Pennsylvania 1795-'96, Speaker of the Pennsylvania State Senate 1801-1803 and presided at the impeachment trial of Judge Addison. On Oct. 16, 1908, the State of Pennsylvania dedicated a monument to his memory at Lewisburg, Pa.

When selecting holiday or birthday gifts, remember the standard works on the Navy! By interesting landfolk in the deeds of our sailors and warships, you render the Navy a distinct service.

NAVY BLUE COMPANY
GREENLAWN, N. Y.

BY EDGAR STANTON MACLAY.

THE STORY OF A HISTORY.

We have often been asked: "How came Edgar Stanton Maclay, a landman, to write so authoritatively on the navy?" The question is natural and in spite of Mr. Maclay's protest against having "my obituary published before I am dead," we will answer it as well as we can.

REV. R. S. MACLAY, D. D.

Although a landman, Mr. Maclay has traveled more extensively on the ocean than many professional sailors besides being an expert in aquatic arts. His father, the late Rev. Robert Samuel Maclay, D. D., was sent out from the Baltimore Conference of the Methodist Episcopal Church in 1847 as one of the first missionaries of that denomination to China. The historian's mother was Henrietta Caroline Sperry, of Bristol, Connecticut.

Edgar Stanton Maclay, the youngest of eight children, was born in Foochow, China, April 18, 1863. When five years old he came to America with his mother where he remained five years and in 1873 went with the family to Yokohama, Japan, remaining there until 1880. It was there that our historian received his first inspiration for writing his History of the United States Navy. His only playmates, during the seven years he lived in the Island Empire, were English, French and German boys; and many were the wordy battles waged between them over the relative merits of their respective countries. Yokohama, at that time, was one of the principal stations for men-of-war of nations maintaining a naval force in the Orient. Being an expert boatman, young Maclay spent many of his leisure hours in his boat on the broad waters of Yokohama Bay, visiting and inspecting American, European and Japanese warships.

Naturally the presence of these belligerent craft provoked discussions between young Maclay and his European playmates. Maclay, while sturdily upholding the American end of these "verbal actions," often found himself at great disadvantage because there was no history of the

THE STORY OF A HISTORY.

United States Navy, at that time, from which to replenish his stock of wordy ammunition, while his antagonists (particularly the English boys) fired broadsides of naval history at him from European historical arsenals. Possibly the most effective discharges came from James' History of the British Navy, written by one who was especially bitter toward Americans. Young Maclay keenly felt his disadvantage and he determined that his first mission on getting out of college (for which his mother all this time had been, personally, preparing him) would be to write a history of the United States Navy.

EDGAR STANTON MACLAY.

In 1880 young Maclay came to the United States and in the following year entered the classical course in Syracuse University, N. Y., being graduated in 1885 with the degree of A. B. Completing his course two months ahead of his class, Maclay went to Europe to gather original material for his naval history. Being supplied only with modest funds by an elder brother, Maclay took steerage passage, which he always declared was a most valuable experience as it placed him on terms of intimacy with the sailors and gave him a practical insight into modern seamanship which has manifested itself so remarkably in all his nautical writings. In his persistency in getting *original* material, Maclay crossed the Atlantic four times, three of the voyages being in the steerage.

He spent more than a year in England, France, Holland and Germany and was fortunate in unearthing much valuable *new* information on American maritime history. In the French Marine and Colonial archives and in the *Bibliothèque Nationale* in Paris he discovered the original official reports of the French commanders concerned in our war against France, 1798-1800, which threw a flood of light on what, down to that time, had been an almost unknown chapter in our national history. In the British Museum Library and in the Admiralty Office in London, where he spent more than six months, Maclay unearthed many new facts bearing on his theme. He expresses most satisfaction, however, over the documents he secured from Sir Provo Wallis in 1886, then the venerable senior admiral of the royal navy, who served as first lieutenant in the English frigate *Shannon* which captured the American frigate *Chesapeake,* June 1, 1813, off Boston harbor. It was this meeting with Sir Provo that Maclay regarded as "spanning a century in historical research." Sir Provo died on February 13, 1892, at the age of a little more than one hundred years.

From Sir Provo, Mr. Maclay obtained proof that settled the theretofore mooted question as to whether or not Captain James Lawrence gave expression to the words "Don't give up the ship"—Sir Provo saying to Mr. Maclay "We heard that when they were carrying Captain Lawrence below, mortally wounded, he uttered the words 'Don't give up the ship.'"

Mr. Maclay also obtained from Sir Provo documents showing that Captain Broke's official report of the *Chesapeake-Shannon* action was an absolute forgery—Broke commanding the latter ship on that occasion. This point was of value in view of the charge that many of the official reports of British commanders of the naval war of 1812, as given to the public, were garbled and misleading. This charge became more serious when, in response to Maclay's request to see the original papers, the Admiralty wrote to him: "their Lordships express to you their regrets at not being able to comply with this request, as the regulations in force preclude all public inspection of admiralty records after the year 1800."

Returning to America in July, 1886, Maclay in two years completed his History of the Navy, then in two volumes, 1775-1866. For four years after that the manuscript went the rounds of nearly every leading publishing house in New York City. Not one would accept it. Some said "We do not see that the market calls for a history of the navy" and yet, at that time no complete narrative of our navy's career was in existence—Cooper's work, which came down only to the Mexican war, long since having been out of print. "The fact of the matter," said Maclay, "was that it was rather too presumptuous for an unknown author, scarce twenty-six years old, to undertake such a pretentious work—and the publishers, very properly looking at it from a strictly business viewpoint, may have been right. Anyway, the manuscript was knocked about in an old satchel for six years, and more than once I was tempted to burn it up."

Determined to emerge from the class of "unknowns," Maclay applied himself diligently to other historical works. He wrote a History of the Maclays of Lurgan (published 1888) which has been pronounced a model of genealogical record. He traced his clan back to the Battle of Bealach nam Broig, 1272, in the Scottish Highlands, and gave a complete account of each of the several hundred descendants of Charles and John Maclay who came to America from Ireland in 1734. The

THE STORY OF A HISTORY.

young historian then published (1890) the Journal of William Maclay (U. S. Senator from Pennsylvania, 1789-1791) which proved to be one of the most important additions to American history in recent years as it gave the only connected narrative of the doings of the present Congress in the first two years of its existence.

Two articles written by Maclay and published in the Century Magazine in the fall of 1890 attracted further attention to the young historian and it was then that several of the publishers who had rejected his manuscript on the history of the navy, re-opened negotiations with him, offering fifty per cent more than the regular royalties. These offers were declined and in 1893 the long-discarded History of the Navy was accepted by D. Appleton & Company and was published in the following year in two volumes. It became an immediate success. A second edition was brought out in 1898 and in the same year appeared Maclay's Reminiscences of the Old Navy, published by G. P. Putnam's Sons. In the following year D. Appleton & Company published Maclay's History of American Privateers, uniform with his History of the Navy.

In 1901 appeared the third edition of Maclay's naval history with the addition of Vol. III covering the Spanish-American War and in 1902 a second edition of Vol. III was issued. In 1903 Maclay published his "Life and Adventures of Jack Philip, Rear-Admiral, U. S. N.," and in the following year "Moses Brown, Captain, U. S. N."

Not the least important of Mr. Maclay's achievements, is his recent discovery of several sea fights in the Revolution which had escaped, for more than a century, all official

OLD FIELD LIGHTHOUSE, WHERE MR. MACLAY WROTE MUCH OF HIS HISTORY.

or formal historical record. Through the courtesy of Prof. Alexander Anderson of the University of Edinburgh, Scotland, and of Dr. Hew

THE STORY OF A HISTORY.

Morrison of the Edinburgh library archives, Mr. Maclay discovered documents showing that hitherto unrecorded American armed craft

MR. MACLAY'S FOUR SONS, WHO ARE "PREPARING" FOR THE NAVY.

had, within two weeks, in March, 1779, attacked four of the enemy's vessels in the Irish Sea.

In was in 1895 that Mr. Maclay resigned from the editorial staff of the New York Tribune and retired into a lonely lighthouse on the shores of Long Island Sound (as lighthouse keeper) in order to concentrate his literary energy on his favorite theme. Here, surrounded by his family, he remained five years. In September, 1900, he secured a transfer to a clerkship in the Brooklyn Navy Yard in order to study more intimately the men and ships of the navy.

On December 23, 1893, Mr. Maclay married and has four sons, each of whom, he declares, will serve in the navy. "As the *personnel* of the navy is now constituted," he said, "with the chance of enlisted men for promotion to the commissioned rank, I cannot think of any profession or mercantile pursuit that offers a more attractive future to the average American youth of refinement, education and good family. Even if the young man does not make it his life work, *a four years' enlistment in our navy is a most valuable supplemental education for the high school or college graduate.*"

THE PUBLISHERS.

GREENLAWN, N. Y., August, 1910.

NAVY BLUE COMPANY
GREENLAWN, N. Y.

THIS BOOK IS DUE ON THE LAST DATE
STAMPED BELOW

AN INITIAL FINE OF 25 CENTS

WILL BE ASSESSED FOR FAILURE TO RETURN
THIS BOOK ON THE DATE DUE. THE PENALTY
WILL INCREASE TO 50 CENTS ON THE FOURTH
DAY AND TO $1.00 ON THE SEVENTH DAY
OVERDUE.

SEP 23 1935	
APR 2 1936	
JUN 4 1941	
5	
JAN 17 195	
	LD 21-100m-7,'33